Brain Theory

Also by Charles T. Wolfe

MONSTERS AND PHILOSOPHY (*editor*, 2005)

THE BODY AS OBJECT AND INSTRUMENT OF KNOWLEDGE
(*edited with Ofer Gal*, 2010)

VITALISM AND THE SCIENTIFIC IMAGE, 1800–2010
(*edited with Sebastian Normandin*, 2013)

Brain Theory

Essays in Critical Neurophilosophy

Edited by

Charles T. Wolfe
Ghent University, Belgium

First published 2014 by
PALGRAVE MACMILLAN

Palgrave Macmillan in the UK is an imprint of Macmillan Publishers Limited,
registered in England, company number 785998, of Houndmills, Basingstoke,
Hampshire RG21 6XS.

Palgrave Macmillan in the US is a division of St Martin's Press LLC,
175 Fifth Avenue, New York, NY 10010.

Palgrave Macmillan is the global academic imprint of the above companies
and has companies and representatives throughout the world.

Palgrave® and Macmillan® are registered trademarks in the United States,
the United Kingdom, Europe and other countries

ISBN: 978–0–230–36957–3

This book is printed on paper suitable for recycling and made from fully
managed and sustained forest sources. Logging, pulping and manufacturing
processes are expected to conform to the environmental regulations of the
country of origin.

A catalogue record for this book is available from the British Library.

A catalog record for this book is available from the Library of Congress.

Transferred to Digital Printing in 2014

Contents

Part III   Evaluation and Speculation

Notes on Contributors

John Bickle is Professor of Philosophy and Adjunct Professor of Psychology at Mississippi State University. He works in the area of philosophy of neuroscience, and is the author of, among numerous publications, *Psychoneural Reduction: The New Wave* (1998), *Philosophy and Neuroscience: A Ruthlessly Reductive Approach* (2003), and most recently, *Engineering the Next Revolution in Neuroscience* (co-authored with Alcino J. Silva and Anthony Landreth, 2013). He has also edited *The Oxford Handbook of Philosophy and Neuroscience* (2009).

Nicolas Bullot is Australian Research Council (ARC) Discovery Research Fellow in Philosophy of Cognitive Science at Macquarie University (Sydney). His research investigates the ability to keep track and identify agents and artifacts over time. His contributions to art theory attempt to bridge the gap between the biological and cognitive sciences of aesthetic appreciation and the historical approach to art behaviors prominent in the humanities and social sciences. His psycho-historical research program for the science of art led to the publication of his collaboration with Rolf Reber as a target article in *Behavioral and Brain Sciences* (vol. 36, "The Artful Mind Meets Art History," 2013). He has received awards from the Fulbright Program (USA), the University of British Columbia (Canada), the CNRS (France), and the Australian Research Council.

Paco Calvo is Chair of the Department of Philosophy at the University of Murcia, Spain. His main area of research is the philosophy of cognitive science. He is the co-editor of the forthcoming *The Architecture of Cognition: Rethinking Fodor and Pylyshyn's Systematicity Challenge* (with John Symons). His articles have been published in *Adaptive Behavior*, *Cognitive Science*, *The British Journal for the Philosophy of Science*, *Mind & Language*, *Minds and Machines*, and *Philosophical Psychology*, among other journals.

Jean-Claude Dupont is Professor of Philosophy and History of Science and a researcher at the CHSSC (Centre d'histoire des sciences, des sociétés et des conflits) at the Université de Picardie Jules Verne (UPJV). His primary areas of research are the history of neuroscience, of embryology, and of pharmacology. He has published, among other books, *Histoire de la neurotransmission* (1999), and recently *L'invention du médicament. Une histoire des théories du remède* (2013).

Luc Faucher is Professor of Philosophy at UQAM (L'Université du Québec à Montréal). He has published widely in the philosophy of cognitive neuroscience, psychiatry, and neuroscience, with edited volumes on emotions, psychopathology, and neurophilosophy. His current project is a monograph on race and racism in analytic philosophy.

Denis Forest is Professor of Philosophy at the University Paris-Ouest, Nanterre, and an associate member of the Institut d'Histoire et de Philosophie des sciences et des techniques (IHPST), Paris. His main area of interest is the impact of the development of brain sciences on our conceptions of human cognitive abilities. He is the author of *Histoire des aphasies* (2005), and he has recently edited a volume on innateness, *L'innéité aujourd'hui* (2013). He is currently writing a book which analyzes various critiques of cognitive neuroscience.

Stephen Gaukroger is Professor of the History of Philosophy and the History of Science at the University of Sydney. His recent monographs are *The Emergence of a Scientific Culture: Science and the Shaping of Modernity, 1210–1685* (2006) and *The Collapse of Mechanism and the Rise of Sensibility: Science and the Shaping of Modernity, 1680–1760* (2010). These are to be continued with a study of the naturalization of the human and the humanization of nature, 1740–1845.

William Hirstein is Professor of Philosophy at Elmhurst College, Illinois. He is both a philosopher and a scientist, having published numerous scientific articles, including works on phantom limbs, autism, consciousness, sociopathy/psychopathy, and the misidentification syndromes. He is the author of several books, including *On the Churchlands* (2004), *Brain Fiction: Self-Deception and the Riddle of Confabulation* (2005), and *Mindmelding: Consciousness, Neuroscience, and the Mind's Privacy* (2012).

Warren Neidich is a Berlin-based interdisciplinary artist and writer. He is recipient of the Fulbright Specialist Program Fellowship, University of Cairo, 2013; the Murray and Vickie Pepper Distinguished Visiting Artist and Scholar Award, Pitzer College, 2012; the Fulbright Specialist Program Fellowship, Fine Arts Category, Faculty of Fine Arts – Ss. Cyril and Methodius University of Skopje, Macedonia, 2011; and the Vilem Flusser Theory Award, Transmediale, Berlin, 2010. Forthcoming publications in 2014 include *Widerstand ist Fruchtbar* and, as editor, *The Psychopathologies of Cognitive Capitalism, Part 2*.

Sarah Robins is Assistant Professor of Philosophy at the University of Kansas. Memory is her primary research interest, but she has related interests in neuroethics, language learning, and tacit knowledge. She has published in both philosophy and psychology.

Sigrid Schmitz currently holds the Chair of Gender Studies at the University of Vienna, and was a visiting professor at the University of Graz/Austria (2003), the Humboldt University of Berlin (2008), and the University of Oldenburg, Germany (2009/2010). Her research and teaching covers approaches in Gender & Science Technology Studies, with particular focus in gender aspects in brain sciences and contemporary neurocultures, body discourses in neo-liberal societal changes, and in feminist epistemologies. Her recent publications include "The Neuro-technological Cerebral Subject: Persistence of Implicit and Explicit Gender Norms in a Network of Change," *Neuroethics* (vol. 5, 2012), Special Issue on *Neuroethics and Gender*, and "Sex, Gender, and the Brain – Biological Determinism versus Socio-cultural Constructivism," in Ineke Klinge & Claudia Wiesemann (eds) *Gender and Sex in Bio-medicine: Theories, Methodologies, Results* (2010).

Katrina Sifferd is Associate Professor in the Philosophy Department at Elmhurst College, Illinois. Her research interests include criminal responsibility and punishment, and the impact scientific theories of decision-making or action should have on culpability. She obtained her PhD in philosophy from King's College, University of London. After leaving King's, Katrina held a post-doctoral position as Rockefeller Fellow in Law and Public Policy and Visiting Professor at Dartmouth College. She is the author of numerous articles and book chapters, including "In Defense of the Use of Commonsense Psychology in the Criminal Law," *Law and Philosophy* (vol. 25, 2006), "The Legal Self: Executive Process and Legal Theory," *Consciousness and Cognition* (vol. 20, co-authored with William Hirstein, 2011), "Translating Scientific Evidence into the Language of the Folk: Executive Function as Capacity-Responsibility," in Nicole A. Vincent (ed.) *Legal Responsibility and Neuroscience*, OUP series on neuroscience and law (2012), and "On the Criminal Culpability of Successful and Unsuccessful Psychopaths," *Neuroethics* (vol. 6, co-authored with William Hirstein, 2013).

Jacqueline Sullivan is Assistant Professor in the Department of Philosophy, Member of the Rotman Institute of Philosophy, and Associate Member of the Brain and Mind Institute at Western University, Ontario. She works primarily in the areas of philosophy of neuroscience and empirically informed philosophy of mind. She is the author of multiple recent journal articles that have appeared in *Philosophy of Science* and *Synthese*. She is also co-editor of the forthcoming *Classifying Psychopathology: Mental Kinds and Natural Kinds* (with Harold Kincaid).

John Sutton is Professor of Cognitive Science at Macquarie University, Sydney. His current research and papers address autobiographical and social memory, distributed cognitive ecologies, skilled movement and embodied expertise, and cognitive history. He is the author of *Philosophy and Memory Traces: Descartes to Connectionism* (1998), and co-editor of *Descartes' Natural Philosophy* (with S. Gaukroger and John Schuster, 2000), a forthcoming volume of essays *Embodied Cognition and Shakespeare's Theatre: The Early Modern Body-Mind* (with Laurie Johnson and Evelyn Tribble), and the international journal *Memory Studies*.

John Symons is Professor of Philosophy at the University of Kansas. He is currently working on topics related to computation, complexity, and emergence. He is the author of *On Dennett* (2001), *Dennett: Un naturalisme en chantier* (2005), and the forthcoming *Theories of Brain Function and the Nature of Vision*. He has edited numerous books and has published many articles in epistemology, philosophy of psychology, and philosophy of science.

Kellie Williamson is a PhD candidate in Cognitive Science and Philosophy at Macquarie University, Sydney. Her current research interests are embodied forms of memory, skilled movement, distributed cognition and emotion experience in performance contexts. She is the coauthor of papers appearing in *Educational Philosophy and Theory* and the *Routledge Handbook of Embodied Cognition*.

Charles T. Wolfe is a research fellow in the Department of Philosophy and Moral Sciences and Sarton Centre for History of Science, Ghent University, and an associate member of the Unit for History and Philosophy of Science at the University of Sydney. He works primarily in history and philosophy of the early modern life sciences, with a particular interest in materialism and vitalism. His current project is a monograph on the conceptual foundations of Enlightenment vitalism. He is the editor of *Monsters and Philosophy* (2005) and co-editor of *The Body as Object and Instrument of Knowledge* (with O. Gal, 2010), and *Vitalism and the Scientific Image in Post-Enlightenment Life-Science* (with S. Normandin, 2013), and co-edited a special issue of *History and Philosophy of the Life Sciences* on the concept of organism (with P. Huneman, 2010).

Introduction

Charles T. Wolfe

> I don't pretend to account for the Functions of the Brain. I
> never heard of a System or a Philosophy that could do it.
>
> (Mandeville 1730, 137)

The present volume is the result of a feeling of dissatisfaction with
current 'discourses of the brain,' from the by-now classic project of
neurophilosophy (as discussed and partially defined below) to more
recent revisions such as neurophenomenology, embodied cognitive
science, but also more theoretical projects such as critical neuroscience,
and the welter of empirical 'neuro'-boosted fields that have emerged in
recent years, such as neuroethics, neurolaw, and, differently, neurofeminism.* Some of these discourses are, of course, featured here, as well as
more historical and evaluative contributions, in the name of conceptual, empirical, and methodological pluralism. The volume has, to be
sure, no claim to offering some kind of exhaustive, synoptic overview
of an entire field – for indeed, there *is no* field. That is, when neuroscientists from Roger Sperry and John Eccles to Jean-Pierre Changeux
and Gerald Edelman wax philosophical, it is not as if their theoretical
terms are clearly demarcated and can be transferred or treated cumu-
latively between their various inquiries, any more than when Patricia
Churchland, Andy Clark, Evan Thompson, or Cordelia Fine address
issues in cognitive architecture, embodiment, or social discourse on
brains.[1] In addition, "the question of what counts as a good explanation
of cognition has not been settled decisively."[2]

Brain theory as presented in various forms here, then, is a looser analytic
category than philosophy of neuroscience, neurophilosophy, or the
more recent 'neurophenomenology,' while at the same time being more
philosophically committed than projects like 'critical neuroscience.' The

former projects tend to take the form of foundational reflection on technical issues in neuroscience (Bickle and Hardcastle 2012), i.e., neurophilosophy seeks to use scientific developments to answer philosophical questions, such as those perennial favorites, consciousness or free will, whereas philosophy of neuroscience takes up more conceptually streamlined items from recent neuroscience in order to continue to do philosophy of science. Now, these approaches no longer have a monopoly on how philosophers understand the issues. Almost wholly opposed views, which stress the irreducibility of 'embodiment' and a role – indeed, a key role – for the phenomenological tradition, in understanding the complex of intentionality, action, and motricity, have also enjoyed prominence in recent years, under the heading of 'enactivism' (Gallagher and Zahavi 2008; Noë 2004; and for more technical versions of the same programmatic view, Petitot, Roy, and Varela, eds, 1999; Thompson 2007). Some go so far as to claim that the neuroscience of action (Jeannerod 1997; Berthoz and Petit 2008) 'confirms' what the phenomenology of embodiment, from the later Husserl to Merleau-Ponty, has insisted on (Petit 1997, 6). But this particular polarity is no longer so prominent in the field, just as, similarly, embodied and situated cognition are no longer exotic (Clark 2008; Shapiro 2010; Menary, ed., 2010; Radman, ed., 2013). In this volume, embodied cognition crops up in unexpected places – or at least, concerning topics rather refreshingly different from 'le corps propre' and enactivism. It is most explicitly present in the chapters by John Symons and Paco Calvo (on embodied robotics) and by Kellie Williamson and John Sutton (on embodied collaboration, thus bringing together embodiment and social cognition).

On the one hand, sympathetic critics have reproached the enactivist theory for its essentialism with regard to selfhood and the first-person perspective (Di Paolo 2009); on the other hand, researchers coming from very different intellectual traditions have sought to integrate select phenomenological insights into their own, naturalistic and reduction-friendly projects (Bickle and Ellis 2005). The present volume is much less concerned with the specific project of brokering a mutually beneficial conceptual outcome for phenomenology and science, or with the goal of somehow rescuing or legitimating the 'enactivist' project (as in Thompson 2007). As mentioned above, it is more pluralistic. That said, the word 'theory' in the title is meant seriously, in the following sense. While there are introductions to 'neurophilosophy' in the classic, Churchlandian sense (e.g. Churchland 2002), they feature none of the more 'critical' or evaluative readings proposed in this volume, whether they concern enactivism, social neuroscience or more historically informed accounts of neuroscientific debates (consider the perspective on memory that emerges from a consideration of Jean-Claude Dupont's chapter on memory traces as a historical problem in neuroscience, together with Sarah Robins' more contemporary reflections on the flaws of 'neural lie detection' and 'brain fingerprinting,' precisely because of their reliance on a discredited, archival model of memory; Robins' intimations of a more 'collaborative' dimension of memory also resonate with Williamson and Sutton's chapter).

Yet the present chapters in brain theory are, in the majority, philosophical (most contributors belong to philosophy departments except for Sigrid Schmitz, Nicolas Bullot and Warren Neidich, who are respectively in Gender Studies, Cognitive Science, or work as independent artist-theoreticians), rather than social and cultural, studies of our relation to the figure of the brain, the popularity of neural imaging, etc. (which is not to say that the philosopher does not have a lot to learn from such studies, e.g. Dumit 2004, Alač 2008). It encompasses foundational, conceptual inquiries (e.g. Stephen Gaukroger's chapter on pain and phantom limb syndrome); empirically motivated projects (e.g. William Hirstein and Katrina Sifferd's contribution on the prospects for 'neurolaw' in the case of psychopathy); as well as more programmatic ones (Jacqueline Sullivan's suggestion for how neuroscience might rethink the call for a 'Decade of the Mind,' Nicolas Bullot's proposed 'psycho-historical' theory of art, neither reductively neuroaesthetic nor merely historicist and relativist); and more speculative chapters such as Warren Neidich's attempt to theorize what he calls 'neuropower,' in between brain plasticity, aesthetic practice, and sociopolitical forms of control (like a more politicized version of what Malafouris calls metaplasticity, i.e. the interpenetration of cultural plasticity and neural plasticity: Malafouris 2010, 267; compare the discussion of cerebral development in Denis Forest's chapter in this volume, dealing with 'neuroconstructivism'). At the same time, some of the chapters here *are* contributions to the progress of neurophilosophy (John Bickle's look at how the 'little e eliminativism' inherent in molecular neuroscience might serve as a constraint on 'neuro-normativity'; Luc Faucher's suggestions for how to improve the social cognitive neuroscience of racial prejudice), or to critical neuroscience (Sigrid Schmitz's examination of the gendered dimension in 'neurocultures').

Despite the presence of the word 'theory' in the title, then, these are not primarily philosophical reflections on neuroscience of the more foundational, a priori sort, exemplified in Bennett and Hacker's 2003 study, which sought to judge all of the science from a purportedly higher, if not transcendental vantage point (John Symons and Paco Calvo in their chapter also discuss attempts like Bennett and Hacker's, to bracket off philosophy from experimental investigation into brain function). We should distinguish foundationalist from empirical approaches here. Foundationalist approaches tend to prefer a priori pronouncements on what the mind is like and why it cannot be identical with the brain. These can be Wittgensteinian, Husserlian, inspired by agent causation or various other sources. A good example is Norman Malcolm's confident assertion (contra the materialist David Armstrong) that "Since intending, hoping, knowing, etc. do *not* have genuine duration, and physico-chemical brain states *do* have it, then intending, knowing, etc. are *not* brain-states" (Malcolm, in Armstrong and Malcolm 1984, 86). Indeed, as Jean-Claude Dupont observes in his contribution, brain theory is not concerned with philosophical positions in which cerebral processes are categorically irrelevant to the understanding of mental life. Stephen Gaukroger notes here that the turn away from purely conceptual, a priori considerations of the nature of mental life (specifically, the experience of pain) is not actually a post-Quinean innovation: Descartes himself, often presented as the chief culprit qua 'inventor' of the modern mind-body problem, was in fact someone who primarily started from physiology, thus empirically, to work his way into such problems: it is by now widely accepted in Cartesian scholarship, not least due to the efforts of scholars such as Gaukroger and Sutton, also a contributor to this volume, that we need to replace our picture of the 'metaphysical substance dualist' Descartes with that of a figure who was primarily a natural philosopher, and one who was explicitly concerned with embodiment, the passions, and the *union* of body and soul (see for instance Gaukroger, Schuster, and Sutton, eds, 2000).

Indeed, there has been something of an 'empirical turn,' also manifest in a number of the contributions to this volume. It can take different forms, not all of which are complementary or in agreement with one another. One, which is very much connected to the emergence of philosophy of psychology as a discipline (and the way it no longer downplayed the significance of biological structures and constraints in the development of psychological theories), is a shift away from metaphysical concerns (physicalism, supervenience, emergence…) to more concrete matters, that can derive from robotics, the study of skilled behavior (including dance and sport, as described here by Williamson and Sutton), or cognitive archaeology.[3] Another, as mentioned briefly with respect to neurophenomenology and enactivism, is the way concepts such as embodiment are now investigated much more in concrete, indeed *embedded* contexts, whether biological (as in Bickle's usage of molecular-level neurobiology), behavioral (as in Sullivan's usage of the Morris water maze experiment) or artificial (as in the usage of robotics in Symons and Calvo's chapter). One thinks also of the way philosophers concerned with the 'systematicity' of human cognition rejected connectionist explanations (according to which the mind can be understood in terms of an interconnected network of simple mechanisms, as in neural networks; connectionists hold that cognitive and behavioral properties can be modeled and explained in terms of their emergence from the collective behavior of simple interacting and adaptive mechanisms) for their lack of relation to this systematic character, in favor of a detailed account of the properties of the neural substrates (Fodor and Pylyshyn 1988, discussed in Calvo and Symons, eds, 2014; see also Symons and Calvo, eds, 2009, xx–xxi).

But the empirical turn is not a panacea. Some object to the turn away 'from ontology to the laboratory' that it leaves philosophical issues unresolved (see also Faucher's friendly critique of Bickle in his contribution). Others, represented in this volume by Sigrid Schmitz, stress that when dealing with any of the core aspects of human agency and personhood, there is a need for social and political critique, unless we uncritically accept, not even neuroscientific evidence, but often, ideological packaging and manufacturing of that 'evidence' (including in the case of 'neural lie detection' studied by Sarah Robins in her chapter). Schmitz's expression for this is "neurocultures": these "set out to explain and predict all modes of thinking and acting of the cerebral subject based on its brain biology" (Schmitz, this volume; see also Farah 2010).

This type of critique is primarily known under the name of 'critical neuroscience' (Choudhury and Slaby, eds, 2012; I note that I thought of the subtitle 'critical neurophilosophy' before becoming aware of the interesting literature in critical neuroscience). As its name indicates, the critical neuroscience program aims in part to criticize current developments, particularly in cognitive neuroscience (Choudhury, Nagel, and Slaby 2009, 73). This can include the already-familiar social critique of our fascination with brain imaging (but also methodological problems inherent in fMRI analyses), the newer critique of 'brain-centric' explanations of personhood, agency, moral life etc., and also, scientifically informed challenges to exaggerated and otherwise ideological reports of neuroscientific findings in popular media (including in the neuropolitical sphere, as discussed below), but also in fields such as the 'neurohumanities.' Just as we are often confronted with bogus neuroscientific explanations (or 'aids') in political decision-making or religious belief, similarly, certain current forms of neuroaesthetic discourse will seek to augment literary scholarship by telling us that in reading literary prose, "the line 'He had leathery hands' has just stimulated your sensory cortex in a way 'he had rough hands' can never hope to" (Walter 2012). Nicolas Bullot and Warren Neidich, in this volume, present a very different case for the relation between current neuroscience and aesthetic practice! (Cultural plasticity need not imply the converse, that a future neuroscience would enable us to 'explain' cultural forms.)

A different type of difficulty with the empirical turn, for the philosopher (here, particularly as regards 'neuroethics,' the law, and other social and moral concerns such as retribution and punishment, the prediction of crime, and so on: what Bickle in his contribution calls the disciplines of "neuro-normativity"), concerns the rise of the 'neuro'-disciplines. The prefix 'neuro-' has become ubiquitous in numerous scientific and loosely scientific disciplines, offering as it does a surplus of concrete, supposedly experimentally substantiated 'brain explanations' for various hotly debated phenomena (from punishment and free will to gender and economic decision-making). But as Jan De Vos has observed, this trend has led to a doubly unfortunate effect: the weakening of the relation of any of these projects to actual neuroscience, and the weakening of the discipline of which they are the 'neuro' version (De Vos forthcoming; see also Ortega and Vidal, eds, 2011). De Vos quotes Matthew Taylor, a British Labour Party activist and government adviser under Tony Blair, who claimed that insights from neurological research offered a more solid base "than previous attempts to move beyond left and right" (Taylor 2009). To the 1980s-type fascination with 'my brain is my self,' the last decade has responded with a particularly vacuous version of a social turn, conveyed in a variety of expressions, from 'neurocapitalism' and 'neuropolitics' to the possibility of neuroenhanced individuals possessing a 'neurocompetitive advantage' (Lynch 2004; Schmitz in this volume).

One problem would be the potentially illusory character of such promised developments. But another problem is in a sense the exact opposite, namely, if neuroenhancement is real, what about "the freedom to remain unenhanced" in a context where schools, in a country we don't need to name, are coercing parents to medicate their children for attention dysfunction (Farah 2005, 37)? The ethical issues here are varied: if students have enhanced mental skills when stimulated by Ritalin, are they cheating? Or (in an example given in Bickle and Hardcastle 2012), treatments for dementia will most likely lead to drugs that increase mnemonic recollection or recall in normal brains as well: would using this drug cross an ethical line from acceptable medical treatments to unacceptable cognitive enhancements if given to members of the general population? Of course, Bickle would not say this is a problem with the empirical turn, on the contrary: the idea is rather that old ethical problems are given a new urgency and a new dimension with changes in neuroscience and pharmacology, which were not present in classic philosophical thought experiments and fictional scenarios. This is also the approach taken by Robins in her contribution. An even stronger embrace of, specifically, 'neurolaw' is in Hirstein and Sifferd's chapter on "the significance of psychopaths for ethical and legal reasoning." If positron emission tomography (PET) studies have already shown that some convicted murderers have significantly attenuated functioning in their prefrontal cortex (a region known to be involved in cognitive control and planning), it is an open book for jurists to plead attenuated responsibility in terms of prior cerebral dispositions. But they take the reasoning one step further, focusing on the specific case of psychopaths and their diminished sense of moral empathy or responsibility. Hirstein and Sifferd effectively argue that the courts need to be practicing 'neurolaw' in order to monitor psychopathic prisoners more closely. Somewhere here there is also the danger of so-called 'brain-realism.' As per Dumit 2003 (see also De Vos forthcoming; Schmitz in this volume), our society seems to place increased weight on brain data compared with other kinds of data; experimental philosophers have shown this to be in popular science presentations, but here, the concern is that brain scans and other pieces of such information will somehow trump other evidence in legal proceedings (Gordijn and Giordano 2010; discussed in Bickle and Hardcastle 2012).

Of the chapters in this volume that touch on these issues, not all are univocally 'for' or 'against' these disciplines, then. Some are critical on strictly factual grounds, others at a more evaluative level. The contents are divided into three somewhat subjectively defined parts: Part I presents more conceptual and programmatic analyses, Part II more empirically oriented contributions, and Part III what I term 'evaluative' approaches – even though some of the chapters explicitly address, e.g. the relation between conceptual and empirical approaches to a specific issue (be it embodied cognition, collaboration, or neuroethics).

In *Part I*, Jean-Claude Dupont looks at the history and philosophy of the notion of 'memory traces,' showing how an interplay between abstract models of cognition and experimental practice produces fruitful scientific developments on the vexed question of the material nature or basis of memory. (It is perhaps not a coincidence that the other contribution by a French author also includes a historical component: Denis Forest's discussion of Wundt and other 19th-century anticipations of modularity, pro and con.)

Stephen Gaukroger, in "Pain and the Nature of Psychological Attributes," reflects on how to do justice to the intuition that it is people, not their brains or bodies, who are in pain, focusing on cases such as pains in phantom limbs, which, as he notes, have been taken since the formulation of Cartesian mind/body dualism to be especially revealing about the nature, location, and sources of pain in particular and sensation in general.

Jacqueline Sullivan suggests how it might be possible for a 'Decade of the Mind' to be successfully appropriated by neuroscience, that is, not as an insistence that the mind could *not* properly be thusly studied, but on the contrary, an extension and refinement of the neurophilosophical program. She shows through a variety of examples that paradigms intended to investigate complex cognitive functions have gradually been introduced into the cognitive neurobiological literature. Put more bluntly, one of Sullivan's suggestions is that there is no particular innate opposition between the intentional stance and experimental neurological work (a very stimulating idea, which to this writer is faintly reminiscent of Enç's remark in the early 1980s that functional language is present even within the reductive vocabulary of neurophysiology, e.g. 'pain receptors,' hence there is no need to fear that functional concepts will somehow be 'lost' by integrating analyses from the neurophysiological level: Enç 1983).

Denis Forest's analysis of neuroconstructivism as "a developmental turn in cognitive neuroscience?" asks the deceptively straightforward question: what happens when science focuses, not on the structure and functions of a developed brain, but on the developing brain, i.e., looking at how functional specialization and extrinsic properties are specified during brain development, with further suggestions for how this could benefit the philosophy of neuroscience, including the debate over 'mechanisms' and inter-level explanations.

Part II begins with two chapters, each of which touches on embodiment and embodied cognition, although in very different ways, as noted above.

John Symons and Paco Calvo's "Computing with Bodies: Morphology, Function, and Computational Theory" argues that one of the most important constraints on computational theorizing in the study of mind is the initial determination of the challenges that an embodied agent faces. One of the lessons of an area of robotics known as morphological computation is that these challenges are inextricably linked to an understanding of the agent's body and environment, with no need to have recourse to thought experiments.

Kellie Williamson and John Sutton's "Embodied Collaboration in Small Groups" looks at the recent shifts within philosophy of mind and cognitive science, in which many theorists have broadened their questions and practices to focus on the complex and intricate cognitive and affective processes that spread beyond a single individual's brain – distributed across the body and/or the environment, co-opting objects and driving interactions with other individuals. They examine cases of skillful collaborative behavior and shared intentionality notably in cases drawn from sport, illustrating embodied interactions (and embedded, situated, distributed cognition) between team and group members. Again (like Symons and Calvo but also Bickle), Williamson and Sutton's emphasis is on how to build an analysis, not from a purely conceptual investigation but "via theorizing about real world contexts."

In his endearingly titled "Little-e Eliminativism in Mainstream Cellular and Molecular Neuroscience: Tensions for Neuro-Normativity," John Bickle seeks to articulate a kind of eliminativism which would no longer be an ontological thesis (as in classical forms of reductionism) but a thesis "about the actual practices of current neuroscience," where certain molecular mechanisms for cognitive functions turn out to conflict with accepted cognitive-level explanations: this is eliminativism 'with a little e.' Bickle acknowledges that philosophers (and even some cognitive neuroscientists) will find it difficult to skip so many "levels" separating mind from molecular processes, and he describes how many neuroscientists are doing exactly that, in their ongoing laboratory research. In a way that resonates with some of the other chapters (Robins, Schmitz, and differently, Hirstein and Sifferd), Bickle notes that this 'little-e eliminativism' presents challenges to the neuro-disciplines, specifically claims regarding 'neuro-normativitiy,' because neuroethics and the related fields rely directly on cognitive explanations, and thus may not be on solid ground neuroscientifically (compare De Vos, forthcoming).

William Hirstein and Katrina Sifferd, in "Ethics and the Brains of Psychopaths: The Significance of Psychopaths for Ethical and Legal Reasoning," suggest that the emerging neuroscience of psychopathy will have important implications for our attempts to construct an ethical society. In rather clearly normative terms, they advocate a usage of this neuroscience of psychopathy in the courts (among others).

Sarah Robins, in her chapter "Memory Traces, Memory Errors, and the Possibility of Neural Lie Detection," asks a question which to some of us may sound as if it comes from a Philip K. Dick film adaptation: can there be a test for memories of a crime committed, i.e., neural lie detection? In fact, she examines actual resources claiming to provide this feature ("Brain Fingerprinting," a type of neural interrogation of suspects), and shows that they depend on what she calls an Archival View of Memory, according to which the brain stores memory traces of particular past events. Instead she cites support for a 'constructive view of memory' – without, however, discounting outright the possibility that such techniques could be improved.

In *Part III*, Sigrid Schmitz puts forth a feminist analysis of the establishment and transformation of gender norms within neuroscientific research and "neurocultures." She uses the concept of brain plasticity to bring out its potential for deconstructing sex-based neurodeterminisms (like Cordelia Fine, as Schmitz notes), stressing the mutual entanglements of the biological and the social in brain-behavior development. Similar to what is advocated by the critical neuroscience theorists, neurofeminist research axiomatically questions the theoretical underpinnings of the empirical 'evidence' and challenges the interpretation of findings. In an interesting resonance with Neidich's chapter (discussed below), Schmitz takes the discourse of plasticity as a basis for reconstructing the concept of 'brainhood' and the articulation of normative demands for the cerebral subject in a neoliberal society.

Luc Faucher's "Non-Reductive Integration in Social Cognitive Neuroscience: Multiple Systems Model and Situated Concepts" also has embodied or 'situated' cognition make an appearance, with the particular case of the social cognitive neuroscience of race (or rather of racially prejudiced cognition). Faucher suggests that this discipline can integrate insights from the aforementioned theories. He additionally points to how this field combines reductionist strategies, such as the usage of brain mapping studies, with an attempt to articulate a more heuristic model, which draws on a "socially situated and embedded theory of concepts."

Nicolas Bullot, in "History, Traces, and the Cognitive Neuroscience of Art: Specifying the Principles of a Psycho-Historical Theory of Art," proposes a ("psycho-historical") theory of art, partly inspired by Dennett's 'intentional stance,' which is meant to avoid the twin excesses of either radical neuroaesthetics (that is, a purported science which will find universal laws of aesthetic experience) or radical historicism (an exclusive emphasis on historical, interpretive, or cultural context). In a way that resonates with the cognitive archaeology of Malafouris, but also with the more theoretical, speculative propositions of Neidich in this volume, Bullot argues that works of art are necessarily material traces of human agency.

The volume concludes with Warren Neidich's "The Architectonics of the Mind's Eye in the Age of Cognitive Capitalism." In an chapter which brings together his own artistic practice and a number of both established and current neuroscientific themes, Neidich tries to move beyond the glorification of plasticity to a consideration of its socio-political determinants and ramifications, which he calls 'neuro-power.' He defines the latter through a combination of the "re-routing of the long and short-term memories through working memory in the production of future decisions" and as a force acting on "the neural plastic potential of the brain in a living present," especially in the critical periods of development (for more on this aspect see Forest's discussion of neuroconstructivism), with the constant leitmotif of the intent to produce a conscripted and enrolled individual of the future.

The chapters commissioned for this volume seek to broaden the field by dealing not only with neurophilosophical topics (Bickle, Sullivan, Robins, and Hirstein and Sifferd) or more empirical matters (Forest, Williamson and Sutton, Symons and Calvo, the latter two of which give a new twist to the embodied, extended mind debates) but also with pain and personhood (Gaukroger), memory (Dupont, and again Robins), social cognition (Faucher), gender (Schmitz), and neuroaesthetics (Bullot, Neidich). They are written by noted figures in their respective fields, and by bringing them together, a newer and more refined picture of what 'brain theory' might mean emerges.

Notes

*Thanks to John Symons for encouragement early on; to Claudia Alexandra Manta for her efficient assistance; to Sara O'Donnell for many other things.

1. Sperry 1952; Eccles 1970; Changeux 1985; Edelman 1992; Churchland 2002; Thompson 2007; Fine 2010.
2. Symons and Calvo, "Systematicity: An Overview," in Calvo and Symons, eds, 2014.
3. Compare the more abstract discussion of the 'extended mind' in Clark and Chalmers 1998 to that, focusing on tools and technology, found in Malafouris and Renfrew, eds, 2010; Malafouris 2013 (see also Iriki 2009); Radman, ed., 2013. Back in 1999, Clark Glymour, in a critique of Jaegwon Kim's reliance on Davidson and Putnam in building what was meant to be a naturalistic approach to the mind, had warned about the danger of philosophy of mind losing itself in conceptual aporias (Glymour 1999).

References

Alač, M. (2008). Working with Brain Scans: Digital Images and Gestural Interaction in fMRI Laboratory. *Social Studies of Science*, 38, 483–508.

Armstrong, D.M. and Malcolm, N. (1984). *Consciousness and Causality: A Debate on the Nature of Mind*. Oxford: Blackwell.

Bennett, M.R. and Hacker, P.M.S. (2003). *Philosophical Foundations of Neuroscience*. Oxford: Blackwell.

Berthoz, A. and Petit, J.-L. (2008). *The Physiology and Phenomenology of Action*. Oxford: Oxford University Press.

Bickle, J. and Ellis, R. (2005). Phenomenology and Cortical Microstimulation. In D. Woodruff Smith and A. Thomasson, eds, *Phenomenology and the Philosophy of Mind* (pp. 140–163). Oxford: Oxford University Press.

Bickle, J. and Hardcastle, V. Gray (2012). Philosophy of Neuroscience. *eLS*. http://www.els.net (doi: 10.1002/9780470015902.a0024144).

Calvo, P. and Symons, J., eds, (2014). *The Architecture of Cognition: Rethinking Fodor and Pylyshyn's Systematicity Challenge*. Cambridge, Mass.: MIT Press.

Changeux, J.-P. (1985). *Neuronal Man: The Biology of Mind*, trans. L. Garey. New York: Pantheon.

Choudhury, S., Nagel, K., and Slaby, J. (2009). Critical Neuroscience: Linking Neuroscience and Society through Critical Practice. *Biosocieties*, 4(1), 61–77.

Choudhury, S. and Slaby, J., eds, (2012). *Critical Neuroscience: A Handbook of the Social and Cultural Contexts of Neuroscience*. Chichester: Wiley-Blackwell.

Churchland, P.S. (2002). *Brain-Wise: Studies in Neurophilosophy*. Cambridge, Mass.: MIT Press.

Clark, A. (2008). Pressing the Flesh: A Tension in the Study of the Embodied Embedded Mind? *Philosophy* and Phenomenological Research, 76(1), 37–59.

Clark, A. and Chalmers, D. (1998). The Extended Mind. *Analysis*, 58(1), 7–19.

De Vos, J. (forthcoming). The Death and the Resurrection of (Psy)Critique: The Case of Neuroeducation. *Foundations of Science*.

Di Paolo, E. (2009). Extended Life. *Topoi*, 28, 9–21.

Dumit, J. (2003). *Picturing Personhood: Brain Scans and Biomedical Identity*. Princeton: Princeton University Press.

Eccles, J.C. (1970). *Facing Reality: Philosophical Adventures by a Brain Specialist*. New York: Springer.

Edelman, G. (1992). *Bright Air, Brilliant Fire: On the Matter of the Mind*. New York: Basic Books.

Enç, B. (1983). In Defense of the Identity Theory. *The Journal of Philosophy*, 80(5), 279–298.

Farah, M.J. (2005). Neuroethics: The Practical and the Philosophical. *Trends in Cognitive Science*, 9(1), 34–40.

Farah, M.J. (2010). Neuroethics: An Overview, in M.J. Farah, ed., *Neuroethics: An Introduction with Readings* (pp. 1–10). Cambridge, Mass.: MIT Press.

Fine, C. (2010). *Delusions of Gender: How Our Minds, Society, and Neurosexism Create Difference*. New York: Norton.

Fodor, J. and Pylyshyn, Z. (1988). Connectionism and Cognitive Architecture: A Critical Analysis. *Cognition*, 28, 3–71.

Gallagher, S. and Zahavi, D. (2008). *The Phenomenological Mind: An Introduction to Philosophy of Mind and Cognitive Science*. London: Routledge.

Gaukroger, S., Schuster, J.A., and Sutton, J., eds, (2000). *Descartes' Natural Philosophy*. London: Routledge.

Glymour, C. (1999). A Mind Is a Terrible Thing to Waste. *Philosophy of Science*, 66(3), 455–471.

Gordijn, B. and Giordano, J.J. (2010). *Scientific and Philosophical Perspectives in Neuroethics*. Cambridge: Cambridge University Press.

Iriki, A. (2009). Using Tools: The Moment When Mind, Language, and Humanity Emerged. *RIKEN Research Report*, 4:5, online at http://www.rikenresearch.riken. jp/eng/frontline/5850

Jeannerod, M. (1997). *The Cognitive Neuroscience of Action*. Oxford: Blackwell.

Lynch, Z. (2004). Neurotechnology and Society (2010–2060). *Annals of the New York Academy of Science*, 1013, 229–233.

Malafouris, L. (2010) The Brain–Artefact Interface (BAI): A Challenge for Archaeology and Cultural Neuroscience. *Social Cognitive and Affective Neuroscience*, 5(2–3), 264–273.

Malafouris, L. (2013). *How Things Shape the Mind*. Cambridge, Mass.: MIT Press.

Malafouris, L. and Renfrew, C., eds, (2010). *The Cognitive Life of Things: Recasting the Boundaries of the Mind*. Cambridge: McDonald Institute for Archaeological Research Publications.

Mandeville, B. (1730). *A Treatise of the Hypochondriack and Hysterick Diseases, in Three Dialogues* (2nd, revised edn; 1st edn 1711). London: Tonson.

Menary, R., ed., (2010). *The Extended Mind*. Cambridge, Mass.: MIT Press.

Noë, A. (2004). *Action in Perception*. Cambridge, Mass.: MIT Press.

Ortega, F. and Vidal, F., eds, (2011). *Neurocultures: Glimpses into an Expanding Universe*. Frankfurt: Peter Lang.

Petit, J.-L. (1997). Introduction, in Petit, ed., *Les neurosciences et la philosophie de l'action*. Paris: Vrin.

Petitot, J., Roy, J.-M., and Varela, F., eds, (1999). *Naturalizing Phenomenology*. Stanford: Stanford University Press.

Radman, Z., ed., (2013). *The Hand, an Organ of the Mind: What the Manual Tells the Mental*. Cambridge, Mass.: MIT Press.

Shapiro, L. (2010). *Embodied Cognition*. London: Routledge.

Sperry, R.W. (1952). Neurology and the Mind-Brain Problem. *American Scientist*, 40(291), 3–12.

Symons, J. and Calvo, P., eds, (2009). *The Routledge Companion to Philosophy of Psychology*. London: Routledge.

Taylor, M. (2009). Left Brain, Right Brain. *Prospect*, 46 (September). http://www. prospectmagazine.co.uk/magazine/left-brain-right-brain/#.UmWAmfm-2Po

Thompson, E. (2007). *Mind in Life: Biology, Phenomenology, and the Sciences of Mind*. Cambridge, Mass.: Harvard University Press.

Walter, D.G. (2012). What Neuroscience Tells Us About the Art of Fiction. http:// damiengwalter.com/2012/06/10/what-neuroscience-tells-us-about-the-art-of-fiction/

Part I
Concepts and Prospects

1
Memory Traces between Brain Theory and Philosophy

Jean-Claude Dupont

The theory of memory traces and its critics have a long history. From the beginning, long before the first actual characterization of memory traces, memory acquired an ontological status, whether the latter challenged the relevance of traces a priori, or on the contrary legitimized it. Today, the search for memory traces has become a fundamental part of neuroscience. In this context, their significance and their contribution to brain theory is actively debated. The philosophical discussions on the pertinence of research on brain traces continue. For philosophers like David Krell, such debates are like an illusory quest for the Holy Grail (Krell 1990). For others, like John Sutton, there is a healthy continuity between old ideas and contemporary connectionism (Sutton 1998).

I seek to evaluate the scope of philosophical discussions about the status of memory in relation to the history of the memory trace theory. After a brief review of the conditions for the emergence of memory traces as a scientific concept, I focus on the crucial period before the 1960s, a period of development of the epistemologically determinant axes of the philosophy of mind. I hope to put into perspective the current debate on memory traces and to improve the understanding of its evolution, between brain theory and philosophy.

1.1   Memory traces: the origin of a scientific concept

1.1.1   From the metaphor to memory trace

Historians have typically focused on demonstrating the extreme antiquity of the notion of memory trace. The description of memory operations now claims to be free of any metaphor or analogy, as the sole result of the reconstruction of different memory functions and the patient search for their organic "correlates" by means of the most sophisticated techniques. However, the permanence of metaphors, and often their heuristic value, is now well documented (Colville-Stewart 1975, Draaisma 2000).

Based on the Greek idea of *typos*, Western culture early on used two complementary images to express the embodiment of memories: memory as wax tablets on which information is recorded, and memory as store or inventory. With these models, the discourse on memory moved towards increasing 'embodiment' or 'cerebration,' in intention at least, given the lack of instrumentation and operative brain theory. From the Greek metaphors onwards, the idea that memory must be associated with a physical change runs through the entire history of Western thought, from the clinical observations of Renaissance physicians to the cerebral physiology of the Enlightenment.

1.1.2   The contribution of the empirical sciences

For example, the physician David Hartley combined Newton's vibrational theory with Locke's associationist theory of ideas to formulate a theory of memory traces (Hartley 1749). The idea is conceived as a change in the vibrational state of a portion of the nervous system. The "composition" of the waves is the physical correlate of the association. The impression of the waves is that of memory, that assumes the nervous matter to have plastic properties. From the 18th century onwards such attempts to understand memory traces by analogies with chemical or physical phenomena abound: vibration and resonance phenomena, movements or retention of nervous fluid, storage of electrical energy, magnetic storage, hysteresis, elasticity, principle of inertia, crystallization, phosphorescence, and later autocatalytic process and molecular rearrangement. This tradition of understanding psychic memory as a physiological phenomenon often inspired by a physical phenomenon, i.e. organic theories of memory, is variously estimated. David Krell discusses successive forms of physical memory discourse in Plato, Aristotle, Hobbes, Locke, Descartes, Hartley, and Freud, explaining that they all fail to understand the phenomenon (Krell 1990). The appreciation of John Sutton is quite different. He imputes a continuity of the concept of memory traces from Descartes to connectionism, in a sort of triumphal march (Sutton 1998). But even if the permanence of the notion of trace is reflected on, negative or laudatory assessments do not help the historian, who should rather deal with the crucial question of beginnings, of the break or the inflection point that changed the notion of trace as an operational concept. The historian must identify the intellectual and material conditions for the transition from a desired science to an effective science.

1.1.3 From organic memory to memory as a scientific object

These conditions seem to involve primarily the brain sciences: the theory of the neuron, neurophysiology of learning, and recognition of brain plasticity are all steps that gave material substance and cerebral reality to the notion of trace. But what is also important is the notion of a break (*rupture*): the establishment of memory trace in a modern sense needed to reject the idea of a general organic memory. The last quarter of the 19th century saw the birth of the concept of organic memory, at the junction of biology and psychology. This concept of organic memory identified, beyond a simple comparison, mechanisms of hereditary memory and those of mental memory. The extraordinary expansion of organic memory in the scientific and medical literature of that period must be remembered. The Viennese physiologist Ewald Hering best expressed this concept in his famous essay on "Memory as a universal function of organized matter" (Hering 1870) and Richard Semon, who invented the term 'engram,' was its last defender (Semon 1904). It implied the inheritance of acquired characters, but the gradual rejection of this doctrine in favor of Morgano-Mendelian genetics marked the beginning of the end for mnemonic heredity. This impasse was fruitful since it left open the scientific field for the study of human memory.

But this dynamics of the rejection of organic memory is not sufficient to understand the emergence of human memory as a scientific object. It is really in the late 19th century that the science of memory of the individual man begins. In the field of knowledge, the situation of memory changes both in writings and in practice. Human memory could no longer be either a more or less noble element within a general philosophy of mind, or an intellectual property that could be manipulated to improve it. It could not be included, more or less harmoniously, in an elegant but speculative philosophical psychology. Memory must be constituted as an object whose study is in itself a sufficient purpose, that is to say as a true object of science, explored by various emerging disciplines: psychology, psychiatry, neurology. Although specialized writings on memory were previously very numerous, specific strategies to elucidate cerebral memory functions are clearly identifiable only in the 19th century. Thus, emerging experimental psychology, which initially focused mainly on the perceptual issues, organized around memory while psychiatrists developed taxonomies of clinical amnesia, and biologists developed new psychophysiological hypotheses. Frantically observed, experienced, dissected, memory is naturalizing. Whatever the age of the physiological theories of memory and their metaphors, the history of the science of memory, understood as a scientific object worked on by different competing communities, hardly spans one and a half centuries (see Dupont 2005).

1.1.4   From the biological basis of memory to the paradigm of cognitive neuroscience

As well as marking a break, the notion of trace becomes operative only after a convergence which involves much more than brain science. The challenges are formidable: psychology is traversed throughout its history by multiple and recurrent dichotomies (one / multiple, normal / pathological, function / structure, unconscious / conscious), revealing the complexity of the problem of mental functions and especially of memory. The biological notion of trace attempted to integrate these difficulties, to deal with the complexity revealed by psychologists and psychiatrists. It is from this starting-point that we should understand the development of memory trace theories. Displacing even organic analogies (with sensory phenomena, reflexes, hormonal, immunological phenomena, etc.), theorists abandon the illusory hope of identifying specific molecules of memory, turning to specific synaptic models, that is to say true organic theories of memory. This development is only effective in the second half of the 20th century. Eric Kandel's research on the neurobiological substrates of memory is based on the old concept of the plastic neural network and use molecular mechanisms of neurotransmission. This is the direction of the history of the biology of memory: it looks for finer biological correlates of different types of memory revealed by pathology and cognitive neuroscience, and analyzes the possibility of storing in the brain different types of information at cellular and molecular levels. This in compounded by results and some undeniable successes of functional brain imaging over a number of years.

But from a philosophical point of view, the fundamental issues remain as a legacy.

In this regard, neuroscience seems to manage the issues raised by classical philosophy, moving in the direction of traditional solutions, while being more or less aware of the difficulties. The classic mind-body problem had left out some insurmountable complications, at least in a naturalistic perspective. While maintaining two distinct 'substances' left the mystery of their interaction intact (unless one adopted the dualist positions of the electrophysiologist John Eccles (Eccles 1994), parallelism denied interaction without explaining the apparent strict correlation between physical and mental phenomena (now dramatically evidenced by functional brain imaging techniques). While rejecting any form of dualism, cognitive neuroscience has unequivocally chosen the world of representations. Mental phenomena are postulated as a special class of natural phenomena, accessible to natural science and definable by their causes and their effects. The challenge is to show how this causal role is exercised, i.e., to describe how the neurophysiological mechanisms perform these causal functions and how physical phenomena underlie mental phenomena. The mind is a patchwork of emergent properties of brain function. Even the consciousness that accompanies certain mental activities seems to emanate from the brain functioning.

In this perspective, whatever the nuances of the positions taken, the mind keeps its ontological reality, but in some way functionally "disappears" in the brain. To say that all the laws of nature (biological, physical, social) are logical consequences of physical laws (and thus explain inorganic processes) may not eliminate the mind, but amounts in practice to considering the mind as a sub-species of physics, an epiphenomenon. It is therefore true that cognitive neuroscience is often causalist, reductionist, and materialist, and that these positions often seem the least doctrinal and the most natural for biologists committed to the "trace program."

1.2   Memory traces and the history of philosophy of mind

I now suggest a genesis of this unconscious paradigm by following the transformation in the philosophical status of memory, which reflects the historical constituent lines of classical philosophy of mind. Three reductions of memory were discussed in the 20th century: nominalist, behavioral, and neurological. These types of reduction each grant a different ontological status to memory, explaining the decrease of critical positions on memory traces and the increased philosophical legitimacy of the empirical research program on traces.

1.2.1   Nominalist reduction, or memory as a fundamental logical relationship

The reduction of the mind originates in the neopositivist tradition. Memory initially provides a purely logical status, which by definition can have no empirical posterity. The philosophical project of *Aufbau* constitutes the starting point for discussions of physicalism (Carnap

1928, translated in 1967). It involves the determination of the logical forms that were necessary and sufficient conditions for knowledge of the world. For that it was necessary to show the differentiability of objects and concepts from each other, i.e. to translate logical propositions on complex objects into propositions including the symbols of the fundamental objects and logical symbols. The value of the system will depend on the purity of these derivations, just as the value of an axiomatic system in mathematics depends on the purity of the derivation of a theory from the axioms. That of Bertrand Russell and Alfred N. Whitehead in the *Principia Mathematica* inspired the project.

To understand the role of memory in the logical construction of the world, one should remember that the latter faces at least two problems: the choice of the basic elements from which everything else is built, and the choice of basic relations, that is to say, the logical forms by which the shift from one level to another occurs. According to Rudolf Carnap, the basic elements are the elements of experience: it is a phenomenal basis consisting of statements about perceptual experiences. Carnap describes several basic relations: the similarity of parts and the recollection of similarity, referred to as Er (= *Errinerung, ibid.*, § 78). By recollection, Carnap does not intend to introduce the reproduction of an already faded experience, but the retention of an immediately preceding experience, a simple relation to something that happened a short time before, for example a perception. But Carnap cannot and does not want to say more about the nature of what is retained. All he cares about is that any proposition on any object of knowledge can thus be reduced to a proposition on the basic elements, with recollection of similarity as the basic and sufficient relation. Quality classes are the first elements of elementary experiences, that is to say, the qualities of sensations. With the help of these spatio-temporal qualities, the world of physical objects can be built, and then more complex objects can be described, particularly psychological and mental objects. The *Aufbau* is based upon logical and formal aspects of the theory of classes, the theory of relations, and in empirical terms, upon a solipsistic basis (knowledge is primarily subjective). The system of the constitution is formed at the base of a fundamental first relationship, recollection of similarity, and by the first elements of experience. To all that must be added the choice of a basic language: logic. Science does not deal with content but with the form. It is only in this sense that experiences can be seen as the basis of the system. Carnap distinguishes two kinds of descriptions: descriptions of properties and descriptions of relations, and the *Aufbau* only contains the latter. This description of relations is purely structural.

It is clear that under these conditions the elementary experiences differ from qualitative factors in a very psychological way. Elementary experiences have no properties. Propositions on these experiences can be united by their relationships and not by their qualitative determinations. For example, visual and acoustic perceptions are not personal experiences, but are first sorted by their comparison. Similarly, it is clear that memory should not be considered as a function in a psychological sense, but as a logical relationship (recollection of similarity).

1.2.2 Behavioral reduction, or memory as a psychical disposition

The mind-body problem itself can be nothing other than a pseudo-problem. Not only are all philosophical problems dissolved, but the *Aufbau's* nominalist project seemed to discourage the development of empirical sciences such as psychology and neurology. In particular, the search for memory traces was not applicable. Despite Carnap's frequent references, circulation between the logical structure of the world and the world of actual science remained difficult. This is the main reason for the difficulties of phenomenalism, also because of the 'logistic' style of the beginnings of the philosophy of mind.

In order to arrive at basic relations, Carnap started from the constitution of physical objects – this being a way to reconnect with the physical world and thus avoid solipsism.

The debate about basic statements and "protocol sentences" upset the situation by producing the thesis of physicalism. Spurred on by Otto Neurath, Carnap himself gradually pulled back from phenomenalism and adopted a physicalist reductionism.

The problem of translation had to be faced: are all protocol statements translatable into the language of physics? If so, then the language of physics could become the universal language of unified science. In 1932, Carnap began to reduce statements of psychology into the language of physics.[1] According to Carnap, the statement "A is angry" can be translated into statements containing descriptions of observable behavior of A, like "A has a particular symptom of anger." Similarly, "sugar is soluble in water" means "in the water, it would dissolve"; "yesterday I was angry" means "yesterday, my body had the symptoms of anger." "Sugar is soluble in water" is the same from a formal point of view as "yesterday I was hungry," as both statements indicate a disposition: in such circumstances, such events occur. Mental concepts allow of such reconstructive logical analysis, aimed at establishing necessary and sufficient behavioral conditions for their application. With dispositional terms, Carnap thinks he can translate any language in the psychology of behavior, which is itself reducible to the physical. Protocol statements are reduced to some publicly verifiable and revisable physical statements, and solipsism seems to be avoided.

Naturally, Carnap's physicalism excludes the world of psychical representations. Can it be said that all these physical symptoms express a mental state without identifying with it? It is impossible because the proposition "There is a mental state independent of its manifestations" cannot be verified empirically. The proposal is untestable, therefore nonsense, a metaphysical pseudo-proposition. Two different languages, a psychic language and a body language, are available to us to express the same theoretical content. Even if the psychic language expresses more things, they are only concurrent representations that can be excluded from the scope of psychology.

It is therefore understandable that behaviorism (mainstream psychology during this period), sought to build on logical positivism (mainstream philosophy of science). This objective alliance was possible, despite considerable differences concerning the conception of science. Nevertheless, according to both psychological and philosophical versions of behaviorism, behavioral reduction applies itself to all mental states, including memory.

John Watson, quoted by Carnap in the *Aufbau*, republished his book on behaviorism in 1930 (Watson 1930), developing as usual the idea that most behavior can be explained in terms of learning. So memory plays a central role in the founding of scientific behaviorism. His views do not differ significantly from those that point to Vladimir Betchterev, Ivan Pavlov, or Edward Thorndike.

By memory, Watson says, we simply mean the following fact: when a stimulus is presented to us after a certain time, we react by the old habit learned when the stimulus was presented for the first time. For Watson, psychologists should focus on behaviors associated with memory and not on the 'sense of familiarity' that is the hallmark of the consciousness of memory. Admittedly memory exists, but there is no need to postulate mental entities: simply observe how the man acts when he remembers.

Even the term 'memory' is rejected as mentalist. Events close in time and space are associated. Memory is the result of associative links between stimuli and responses. The organism acquires the habit of reacting to environmental stimuli. These habits can be verbal, as in the association of a person's name with a certain person, or manual, such as shaking hands with that person, or even visceral, as when the sight of some people gives you nausea. We are 'organized' by the environment to respond in a variety of ways to different stimuli: nothing is 'stored' in any form. The organization is structured as a whole to show behaviors that are described as memory.

It is clear that behaviorism neither wants to nor can say something else about memory. The current marginalization of behaviorism should not obscure the heuristic value of behavioral reduction. Although the concept of memory is not explicitly developed, the concepts developed by neobehaviorists which were mainstream in the 1930s and 1940s, such as reinforcement theory (Clark Hull), contiguity theory (Edwin Guthries), and the law of effect (Burrhus F. Skinner), greatly stimulated research on learning. From this perspective, however, the search for memory traces is, if not inconceivable, then not relevant.

1.2.3   Neurological reduction, or memory identified with memory traces

But behaviorism experienced major difficulties: assumption of unob-servable, indefinite heterogeneity of behavioral dispositions, implicit intentionality of behavioral categories, overestimation of third-person knowledge, proximity with the neopositivist conception of science, implicit commitment to ontological positions.... Some of these diffi-culties were taken up by Herbert Feigl, thereby sowing the seeds of the identity theme.

Feigl completely renewed discussions of physicalism in two articles: "Physicalism and the Unity of Science," written in 1954, and "The 'Mental' and the 'Physical'" (Feigl 1963, Feigl 1958). With Feigl, trace theory finds an unbelievable support since memory is identified as its traces, paving the way for a brain theory.

The genesis of Feigl's positions is well known. He was one of the first emigrants of the Vienna Circle to settle in the United States. By 1939, Feigl had sought to give biology, and especially neurophysiology, the status of science in a physicalist sense. He gradually broke with Carnap's work on the language of science and abandoned the assumption of a possible translation of psychological language into the language of physics involving only purely analytic rules. Several attempts finally led him to write his 1958 essay, which can be regarded as a return to empiri-cism, both physicalist and realist.

According to Feigl, one must postulate mental states. There are indeed conscious episodic experiences, 'recalcitrant' mental concepts, and residual sensory qualities ("raw feels") that I have direct experience of and that are obviously not simple verbal substitutes of behavior. When someone says, 'I have a pain,' they are not only expressing the verbal behavior of pain. What is reported is something real: the pain. Similarly, dispositions are recalcitrant. Feigl's criticism with regard to the previous model of reduction is the positivist oblivion of the real object, in favor of proofs, and the too-quick relegation of the mind-body problem to the rank of a pseudo-problem.

The objection to realism could be the basis for a return to dualism. But for Feigl, an analysis of 'something' that would involve mental states is not a return to dualism if this 'something' is *identified* with brain states. Based on Frege's classic distinction between sense and reference, Feigl argues that there is a systematic identity, where any psychological predicate denotes an entity corresponding to a physiological predicate. This is what makes the identity non-trivial. 'I see the morning star' and 'I see the evening star' are not synonymous, but the celestial body is the same. The synonymy of expressions on both sides of the 'equal' sign relates only to the reference. The purpose of the reduction is not an identification of concepts, but of references. It is then possible to identify some of the physical referents of concepts with the referents of terms that express the raw feels. It is indeed a monism, since different types of speech denote the same object. However, according to Feigl, these forms of discourse describe different levels of reality. These levels must be identified in a synthetic way connected with the experience and without logical necessity: the level of immediate perception of mental states, the level of psychological concepts, and that of neurophysiological concepts. Feigl prefers this solution to the multiple languages that would describe aspects of the same reality and to the task of translation without epistemology.

Feigl's neurological reduction replaces behaviorist reduction, which contends that psychological discourse is reducible and must give way to dispositions. Similarly, it is clear that eventually only the language of neurophysiology may subsist.[2] Feigl has some startling suggestions due to his confidence in the future of neurophysiology[3]; his analysis of memory research in 1958 reflected his optimism, and sometimes lack of prudence. He inventories the pairs of properties traditionally used to differentiate the mental and physical – but memory is a property of both the physical and the mental. Memory, Feigl says, is traditionally regarded as a property of the mind and not a physical property, citing Russell's *Analysis of Mind* in support, where memory is regarded as a characteristic of the mind. However, he adds against Russell, the biologists Ewald Hering and Richard Semon had in their day focused on memory as a general property of organized matter. Memory cannot be used to differentiate the mental from the physical. The inert material itself is endowed with a kind of memory, directly identifiable with memory traces. Feigl refers also to Donald Hebb and reverberating circuits (Hebb 1949). This double reference to Hering and Semon on the one hand, and Hebb on the other hand, as a common argument in favor of identity, suggests the limits of theoretical discussions on reduction: from organic memory to reverberating circuits, practical modalities of identification or reduction may be very different, and they are so in fact. We are dealing on the one hand with a theory of organic memory (Semon), and on the other hand with an organic theory of memory (Hebb). For Hering and Semon, beyond a simple analogy, a physical phenomenon, heredity, and a mental phenomenon, memory, were wrongly identified and reduced to a physical property of matter as a common biological substrate. This is a good example of identification without foundation, which not only did not have heuristic value, but has proved quite disastrous as regards biology. This also demonstrates the significance of the historical inquiry, which only allows the analysis of specific forms of reduction. In each case, a successful reduction can clearly be assessed retrospectively.

The rest of the story is well known: the development of variants of identity and physicalism. According to Feigl, psychological typology was equivalent to neurological typology, and can be reduced by using the appropriate equivalence in language terms. Indifference to neurophysiology is therefore no longer appropriate. While Feigl is still closely intellectually linked to logical behaviorism and the issue of scientific language, other formulations of identity, from U.T. Place, J.J.C. Smart, or David Armstrong, freed themselves from it to a large extent. Mental states became real and causally efficacious states (see Wolfe 2006). From the 1980s onwards, the latest developments in philosophy of mind resonated with progresses in the empirical sciences. They philosophically supported the legitimacy of the search for traces in cognitive neuroscience and reinforced the paradigm described above.

1.3   The renewal of memory traces in contemporary philosophy and its contribution to brain theory

1.3.1   The weight of the legacy of the mind-body problem

Bertrand Russell, when trying to reconcile memory and perception, already noted that the supposed brain mechanisms were entirely hypothetical (Russell 1921). Dozens of years later, when Feigl wrote "The 'Mental' and the 'Physical,'" little real progress had been made in the search for the engram, besides Hebb's model. The problem of memory traces was far from being resolved empirically, which Feigl was quite well aware of: of the ten open scientific questions in his psychophysiology, three concern memory. Lashley's famous skeptical essay on the search for the engram (Lashley 1950) is a more significant indicator of the mindset of researchers than Feigl's optimism.

Philosophical reflections on memory traces were indeed well ahead of effective empirical evidence. Unlike Carnap, Feigl clearly defends the theoretical research program as a possibility. But, even in his time, the empirical results were still modest. Traces were still postulated, and Feigl's philosophical analyses were necessarily of a broad nature.

In addition, there is no absolute incompatibility between dualism and behaviorism as identity with the empirical facts. As Feigl himself points out, to engage in the process of identifying identities is a matter of philosophical interpretation. If the principles of parsimony and simplicity in favor of identity (beyond simple correlation) can be invoked, the economy is mostly speculative and metaphysical, not empirical. It does not eliminate the obligation to establish systematic correlations, although it is not clear what kind of data should be utilized to select the identity. To Feigl, the thesis of identity even requires empirical verification by systematic correlations. The problem is that it seems that no ontological position in the philosophy of mind, none of these reductions, could be empirically validated, as no facts were able to impose a scientific theory, a given epistemology, nor a fortiori an ontology, on their own. In particular, it is difficult to disagree with Smart's position: the theory of psycho-cerebral identity is a scientific theory only to the extent that a psycho-renal or a psycho-cardiac one can be opposed to it (Smart 1970).

However, the need for a theoretical choice should not be called into question. Although the question of the physical-mental relationship is not an empirical one, it seems, contrary to what Carnap believed, that the choice of a phenomenalist, physicalist, or neurological language is not indifferent. More than an inspiration, this choice implies methodological decisions, which constitute the heuristic side of the empirical sciences, that is to say a research program. The example of memory clearly demonstrates that: it is not indifferent to deal empirically with memory as a formal relationship, as a behavioral disposition, or as material traces. Only the latter choice leads to neuroscience. Only the choice of a language leads to bold conjectures in a field, which can be so interesting to refute.

The program of neuroscience is the elucidation of brain function and concerns the materiality of the memory process. However, this research program in neuroscience is not necessarily linked to identity *sensu* Feigl.

In addition, a research program is not limited to the choice of a language. It involves the establishment of a conceptual apparatus and equipment; that of cognitive neuroscience was not yet available in the 1960s.

1.3.2 Memory traces: a new context for an old debate?

It is clear that the development of memory science in the second half of the 20th century was preceded or was made possible by an earlier philosophical gestation, but the influences were not equally decisive. The neopositivist tradition meant to illuminate the configuration of science even though the empirical results on memory traces were still inconclusive. This proximity with the empirical sciences makes it particularly interesting to consider, at least as far as the phenomenological tradition is concerned.

Trace theory is also significant for the philosophy of science, as a case of how interdisciplinary research is built (Hardcastle 1996). These discussions on the notion of trace show how the current work of scientists and philosophers on the question of memory is inseparable, thus illustrating the effective reorganization of science-philosophy links since Carnap and Feigl's time. Trace theory had to undergo major revisions to meet the scientific requisites for memory, to consider new empirical results, to become a part of a new brain model. Scientists tended to refine positions which were not always explicit.

Under these conditions, what can trace theory consist of today and what might be a pertinent critique of memory traces?

John Sutton tries to understand the different strategies developed against the theory of traces (Sutton 1998). Positions critical of the empirical research program develop arguments to show that this research is based on unfounded assumptions, which are unacceptable from a logical, epistemological or ontological point of view. First, there is the phenomenological argument. The diversity of phenomena revealed makes memory unanalyzable and certainly not reducible to a single causal phenomenon such as the trace. Other assumptions could target our assertions based on the physical world, like the assumption of action at a temporal distance: traces are from the past, remembering is from the present. The premise of the traces and the causal theory implies the action of traces at a temporal distance, overcoming the gap between the present and the past. This is inconsistent with the principle of temporal continuity of a cause and its effects (causal contiguity): trace theory might change the notion of causality itself. Another criticism concerns the confusion in the theory of traces between retention and storage, and the fact that trace theory assumes without understanding the isomorphism of traces and representations. Finally, supporters of the traces must resist the idea that there is a sort of internal homunculus who reads or interprets traces, who could link a stored trace with present stimulus, or know how to extract traces according to the circumstances at the time of recall. Such an internal interpreter explains nothing, or leads to the famous regression denounced by Ryle (Ryle 1949). If it is avoided by having further internal mechanisms operate in some 'corporeal studio,' a Rylean regress is generated. Unsupervised mechanisms render the system's contact with the world problematic, and the risk of solipsism follows.

1.3.3   Do memory traces exist?

It is obvious that any theory of memory involves a more general brain theory (and vice versa). Not only are a series of fundamental questions about memory also applicable to the nature of the brain, but memory has generated the development or correction of brain models. Any discussion of memory or one of its aspects involves a representation of the brain and a conception of mind-brain. According to Jonathan K. Foster and Marko Jelicic, the varied viewpoints on memory may be divided into two broad camps:

> the first characterizing memory structurally, i.e. as comprising multiple, different, empirically separable components or systems, and the second conceptualizing memory more in functional terms, as a common, and, some would argue, unitary and indivisible framework. ... While the systems view emphasizes separate neural systems mediating different aspects of mnemonic functioning (e.g. implicit/explicit, episodic/semantic, procedural/declarative), the processing perspective focuses upon the differential mental operations in which individuals engage when confronted by a specific memory task (e.g. deep/shallow processing, conceptual/perceptual processing). (Foster & Jelicic 1999: 3–4)

Even now recent progress in functional brain imaging techniques leaves this question open and attempts at functional connectomics are inchoate (Sporns 2012).[4]

Recent views consider memory and mind as embodied, embedded, and extended. But in what ways might embodiment, action, and dynamics matter for understanding mind and the brain? The challenge to contemporary science is to understand the mind as a controller of embodied and environmentally embedded action. Going on to defend connectionism and dynamic notions of the memory trace, John Sutton asserts that "This more dynamic vision of traces, rejecting the idea of permanent storage of independent items, may satisfy both recent developments in cognitive science and some of the positive suggestions with which critics of static traces have accompanied their objections" (Sutton 2012). He defends the idea that distributed models of connectionism now allow "the dissolution of traditional hostilities between 'direct realist' and 'representationist' approaches to memory" (Sutton 1998: 277).

The notion of trace implies representation. But does cognition need representations? Even if the empirical sciences are now mainly representationist, the concept remains questionable. For Gilbert Ryle, the use of psychological typology was still needed, provided not to confuse the concepts of dispositions and concepts of object, and to place the mental and physical entities in different logical categories. Wittgenstein's critical position concerns both reduction and representations. The language game in which the word memory is used can be explained without postulating internal memory states or traces. According to linguistics or Wittgenstein's 'grammatical' behaviorism, behaviorism was true in the sense that we have only the behavior and language of others to gain access to their interiority, and that is why our discourse on the mental is limited. This "truth" of behaviorism is particularly apparent in Wittgenstein's treatment of memory (Wittgenstein 2009, § 306–307; see Stern 1991).

So the trace is mainly under fire by followers of Wittgenstein (see Malcolm 1977). The criticism of the concept of trace invokes the principle of economy: trace theory and its concepts are not logically necessary. Not only they are not required by the empirical sciences, but also they lead to paradoxes. Certainly neither the theory of indirect representation (representationist) nor direct theory can be empirically established. But the first is illogical, unlike the second, i.e. direct realist (Gibsonian type) that explains the phenomena of memory, but not at the expense of logic. Moreover, the concepts it develops could be inadequate for the analysis of memory. If it can be demonstrated that concepts of isomorphism, encoding, storage, and retrieval are logically inconsistent, no empirical result can save the trace theory (Wilcox & Katz 1981, Sanders 1985). Supporters of traces of course have responded to these fundamental objections by developing other logicist strategies: developing a logical notion of the memory trace, that is to say, by showing that the traces are logically necessary beyond their empirical characterization. Traces have an explanatory virtue, and it is possible to use them for a purely functional purpose. The logical argument brings us back to the logical status of memory as proposed in a very different perspective by Carnap. But it is now clear that the question never will be purely logical.

1.4   Conclusion

The progress of the empirical sciences of memory now allows for more specific reflections. At the turn of the century, bypassing the traditional issues of the mind-body problem, some lines of research are the direct result of the interaction with the empirical sciences. That allows philosophy to fully exercise its function of critical analysis and clarification of concepts.

Empirically, it remains to be proven that the conceptual apparatus and the formalism of neural networks are able to account for what folk psychology realized in terms of memories. Interdisciplinarity implies the need to let live and develop scientific issues, addressing the difficulties inherent in empirical research traces, without immediate theoretical objections. Other philosophical developments in recent years have also been specifically explored, from research on traces to new conceptions of reduction.[5]

So a real interdisciplinary program on the question of traces is being redrawn, where even Wittgensteinian critiques might have positive effects. In fact, only a certain style of philosophical reflection is inoperative here. Brain theory will set aside two types of positions:

1. those who believe a priori that the mechanisms of matter are not philosophical problems, who profess a kind of epistemological dualism of science and philosophy and at the same time, a kind of autonomy of mental life with regard to the biological levels where it emerges. The real question remains rather what would memory be in essence[6];
2. those for whom the essential question, that is to say the epistemological value of memory, is in no way connected to what happens in the brain (Malcolm 1977).

Notes

1. See the two articles published in *Erkenntnis* in 1932 (Carnap 1932a and 1932b) and in *La Revue de Synthèse* (Carnap 1935).
2. He disagrees with Ryle's thesis.

3. For example the fiction of the 'autocerebroscope,' already used by Carnap in *Aufbau* that suggests current techniques of functional brain imaging.
4. See the 1000 Functional Connectomes Project and the Human Connectome Project, described in Sporns 2012, 109–179
5. Schaffner 1999, Bickle 1998, Bechtel 2001, Craver & Darden 2001.
6. See, e.g., Locke 1971 and O'Connor & Carr 1982.

References

Bechtel, W. (2001). The Compatibility of Complex Systems and Reduction: A Case Analysis in Memory Research. *Minds and Machines*, 11, 483–502.

Bickle, J. (1998). *Psychoneural Reduction: The New Wave*. Cambridge, Mass.: MIT Press.

Carnap, R. (1928). *Der Logische Aufbau der Welt*. Berlin: Weltkreis-Verlag; Carnap, R. (1967). *The Logical Structure of the World* (R. A. George, Trans.). London: Routledge & Kegan Paul.

Carnap, R. (1932a). Die physikalische Sprache als Univeralsprache der Wissenschaft. *Erkenntnis*, 2, 432–465. In M. Black (Ed.) (1934), *The Unity of Science*. London: Kegan Paul, Trench, Trubner & Co.

Carnap, R. (1932b). Psychologie in physikalischer Sprache. *Erkenntnis*, 3, 107–142. In A. J. Ayer (Ed.). (1959), *Logical Positivism*. New York: The Free Press.

Carnap, R. (1935). Les concepts psychologiques et les concepts physiques sont-ils foncièrement différents? *Revue de Synthèse*, 10, 43–53.

Colville-Stewart, S. B. (1975). *Physico-Chemical Models of the Memory Storage Process: The Historical Role of Arguments from Analogy*. University of London, PhD Thesis.

Craver, C. F. & Darden, L. (2001). Discovering Mechanisms in Neurobiology: The Case of Spatial Memory. In P. Machamer, R. Grush, & P. MacLaughlin (Eds), *Theory and Method in Neuroscience* (pp. 112–137). Pittsburgh: Pittsburgh University Press.

Draaisma, D. (2000). *Metaphors of Memory*. Cambridge: Cambridge University Press.

Dupont, J.-C. (Ed.). (2005). *Histoires de la mémoire*. Paris: Vuibert.

Eccles, J. C. (1994). *How the Self Controls Its Brain*. New York: Springer Verlag.

Feigl, H. (1958). The "Mental" and the "Physical." In H. Feigl, M. Scriven, & G. Maxwell (Eds), *Minnesota Studies in the Philosophy of Science: Concept, Theories and the Mind-Body Problem* (Vol. 2, pp. 370–497). Minneapolis: University of Minnesota Press.

Feigl, H. (1963). Physicalism, Unity of Science and the Foundations of Psychology. In P. A. Schilpp (Ed.), *The Philosophy of Rudolph Carnap* (pp. 227–267). La Salle, Illinois: Open Court.

Foster, J. K. & M. Jelicic (1999). *Memory: Systems, Process, or Function?* Cambridge, Mass.: MIT Press.

Hardcastle, V. G. (1996). *How to Build a Theory in Cognitive Science*. Albany: State University of New York Press.

Hartley, D. (1749). *Observations on Man, His Frame, His Duty and His Expectations*. London: S. Richardson.

Hebb, D. O. (1949). *The Organisation of Behavior*. New York: J. Wiley.

Hering, E. (1870). Über das Gedächtnis als eine allgemeine Funktion der organisierten Materie. *Almanach der Österreichischen Akademie der Wissenschaften*, 20, 253–278.

Krell, D. F. (1990). *Of Memory, Reminiscence and Writing*. Bloomington: Indiana University Press.

Lashley, K. S. (1950). In Search of the Engram. In *Society of Experimental Biology Symposium, No. 4: Psychological Mechanisms in Animal Behavior* (pp. 454–480). Cambridge: Cambridge University Press.

Locke, D. (1971). *Memory*. London: Macmillan.

Malcolm, N. (1977). *Memory and Mind*. Ithaca: Cornell University Press.

O'Connor, D. J. & Carr, B. (1982). *Introduction to the Theory of Knowledge*. Brighton: Harvester.

Russell, B. (1921). *The Analysis of Mind*. New York: Macmillan.

Ryle, G. (1949). *The Concept of Mind*. Harmondsworth: Penguin.

Sanders, J. T. (1985). Experience, Memory and Intelligence. *The Monist*, 68, 507–521.

Schaffner, K. (1999). Philosophy of Medicine. In M. Salmon, J. Earman & C. Glymour (Eds), *Introduction to the Philosophy of Science* (pp. 310–345). Indianapolis: Hackett Publishing Company.

Semon, R. (1904). *Die Mneme als erhltendes Prinzip im Wechsel des organischen Geschehens*. Leipzig: W. Engelmann.

Smart, J. J. C. (1970). Sensations and Brain Processes. In C. V. Borst (Ed.), *The Mind/Brain Identity Theory* (pp. 52–66). London: Macmillan.

Sporns, O. (2012). *Discovering the Human Connectome*. Cambridge, Mass.: MIT Press.

Stern, D. G. (1991). Models of Memory: Wittgenstein and Cognitive Science. *Philosophical Psychology*, 4(2), 203–217.

Sutton, J. (1998). *Philosophy and Memory Traces: Descartes to Connectionism*. Cambridge: Cambridge University Press.

Sutton, J. (2012). Memory. In Edward N. Zalta (Ed.), *The Stanford Encyclopedia of Philosophy* (Winter 2012 edn), http://plato.stanford.edu/archives/win2012/entries/memory/

Watson, J. B. (1930). *Behaviorism* (2nd edn). Chicago: University of Chicago Press.

Wilcox, S. & Katz, S. (1981). A Direct Realist Alternative to the Traditional Conception of Memory. *Behaviorism*, 9, 227–239.

Wittgenstein, L. (2009). *Philosophical Investigations* (G. E. M. Anscombe, P. S. M. Hacker, and J. Schulte, Trans.). Chichester: Wiley-Blackwell.

Wolfe, C. T. (2006). Un matérialisme désincarné: la théorie de l'identité cerveau-esprit. *Matière première*, 1, 77–100.

2
Pain and the Nature of Psychological Attributes

Stephen Gaukroger

In their *Philosophical Foundations of Neuroscience* (Bennett and Hacker 2003), Bennett and Hacker offer three main arguments regarding the primary bearer of psychological attributes: first, they draw attention to the shift from the Aristotelian idea that the person is the primary bearer of psychological attributes to the Cartesian idea that it is the mind, by contrast with the body, that is the primary bearer; second, they argue that many physicalist projects in the philosophy of mind and neuroscience simply substitute the brain for the mind, and in doing so do not go beyond the Cartesian account in terms of the basic philosophical issues and assumptions; third, they argue for the superiority of an Aristotelian notion of the person as the primary bearer of psychological attributes over the Cartesian (and neo-Cartesian physicalist) view that it is the mind/ brain, as opposed to the body, that is the bearer of these attributes.

In what follows I want to look at one case which has traditionally been taken to show that it is the mind/brain that experiences pain, not the person, and not the body. This is the case of phantom limbs, well known from antiquity onwards, but first used to make a fundamental philosophical point about the nature of pain in Descartes. He notes that one of his critics 'expresses surprise' that he recognizes 'no sensation save that which takes place in the brain'. Yet such a view is, he urges, the only one compatible with medical observation:

> On this point I hope that all doctors and surgeons will help me to persuade him; for they know that those whose limbs have recently been amputated often think they still feel pain in the parts they no longer possess. I once knew a girl who had a serious wound in her hands and had her whole arm amputated because of creeping gangrene. Whenever the surgeon approached her they blindfolded

her eyes so that she would be more tractable, and the place where her arm had been was so covered with bandages that for some weeks she did not know that she had lost it. Meanwhile she complained of feeling various pains in her fingers, wrist, and forearm; and this was obviously due to the condition of the nerves in her arm which formerly led from her brain to those parts of her body. This would certainly not have happened if the feeling or, as he says, sensation of pain occurred outside the brain.[1]

The importance of the brain is that it is the seat of mind, and this is the case whether, like Descartes, one holds that the mind is separate/separable from the brain, or whether one holds that the mind just is the brain. Given the focus of the discussion, brain and mind can be treated as interchangeable terms.

The way in which the phenomenon of phantom limbs is described in both the philosophical and the psychological/neurological literature offers one a choice between accounts of the pain experienced as mental or as physical. But, while there is a sense of 'person' in which it is undeniable that it is the person who experiences the pain, it is assumed that to leave matters here is unilluminating because the further question arises: is the pain in the person's brain or in their body? Indeed, the question seems to have a straightforward but initially surprising answer, for if the question is rephrased in terms of what is necessary for the experience of pain, while there would be general agreement that a mind/brain was necessary, the phantom limb phenomenon seems to show that a body is not necessary. The importance of the phantom limb case is that it confounds our normal untutored assumptions about the location of pains, namely that they are in the limbs in which we experience the pain and to which we apply medical remedies. This way of pursuing the issues seems both eminently reasonable and unavoidable, if we are to throw light on the nature of pain, but once we take this route we seem to be locked into a mind/brain versus body dichotomy, leaving no place for any consideration of the view that it is the person, not their mind/brain, or their body, that is the proper subject of pain: that it is *me* who is in pain, not my mind/brain or my body.

If we are to take this option seriously, we need to investigate what a physiological/neurological account would look like that indicated that the person, not the brain or the body, was the proper bearer of states such as pain. The account in Bennett and Hacker, for example, encounters difficulties with this question in the case of phantom limbs. They argue that pain in phantom limbs is an instance of referred pains, which have two characteristics. First, the location of the cause of the pain differs from the location of the pain itself. Second, it is wrong to say that the sufferer of the referred pain is mistaken about the location of the pain: rather, it is just that the pain is felt in places other than the locus of the injury (Bennett and Hacker 2003, 123–124).

It is certainly true that the location of the cause of the pain and the location of the pain itself differ in the case of referred pain, and the general point that the location of the cause of pain may differ from the location of the pain itself is something that Descartes does not notice, for example. Yet there is a problem for Bennett and Hacker in proceeding in this way. When the person who, having had a heart attack and having experienced a severe pain in his arm, is told by his physician that this is a referred pain, he might naturally interpret this to mean that he had been mistaken in thinking that the pain was in fact in his arm. But presumably one believes oneself mistaken in such cases because, not knowing anything about referred pains, one naturally assumes that the location of the cause of the pain and the location of the pain itself must be identical, so that one thoughtlessly makes an invalid inference from the location of the pain to the location of its cause. But if this is the case, the mistake is an intellectual one, a failure to distinguish cause of pain and location of pain, not a mistaken unconscious or natural judgment about the location of the pain, because the location of the pain has been correctly identified. However, what the intellectual mistake consists in is exactly what the brain/body dualist is pointing to, and indeed generalizing from, maintaining in effect that *all* pains are referred pains, because causes are always (necessarily) located in the brain, whereas they are always (necessarily) experienced in the body. In short, if Bennett and Hacker are right in their claim that phantom pains, like all pains, are experienced by the person, not the person's mind or body, referred pain would seem a distinctly unpromising path to follow in trying to establish this because the prima facie response to a question of location of cause is in terms of the brain (or some part of the brain) versus the body (or some part of the body).

In the circumstances, we might be inclined to be a little more circumspect about generalizing from referred pains, and to treat them as a special kind of case, far from typical of pain in general. Such circumspection is justified on independent grounds, for in this connection, it is worth noting that there are genuine, in the sense of uncontentious, examples of referred sensation in cases of phantom limbs. Ice on the face will routinely elicit cold in phantom digits, for example, and water trickling down the face is sometimes experienced as water trickling down the phantom arm.[2]

Such cases, which are isolated and confined to the face, offer no grounds for moving to the general claim that phantom limb pains are a species of referred pain. Quite the contrary, such exemplary cases, by showing us just how distinctive referred pains are, serve to draw attention to the possibility that the more general suggestion is an inappropriate extrapolation.

2.1   The Proper Subject of Pain

Phantom limb arguments offer a case where we can use the same neurophysiological evidence to compare two different kinds of account of the primary bearers of psychological attributes such as pains: an account that pits mind/brain against the body, and an account that identifies the person as the proper subject of pain. We have just seen, however, that steering the discussion so that the person account even gets to be considered an option is difficult: once one starts probing, the person seems naturally to fall into two parts, mind/brain and body, and everything comes to hinge on the respective contributions of these. How, then, could we possibly test this option against the kind of evidence that is drawn upon in deciding whether it is the brain or the body that is the cause and or locus of pain in the case of phantom limbs? What we need to do, for these purposes, is to translate the person option into something that could be an object of neurological enquiry: we need to identify a neurologically meaningful correlate of the person. This correlate is what we might call an essentially embodied cognitive subject, or essentially embodied brain. This terminology is not wholly satisfactory, but not so important either. What matters is that we are clear about what we are looking for: namely something that necessarily involves a brain/mind that is essentially part of, and functionally inseparable from, the kinds of bodies that we have. If we find that we cannot account in a satisfactory manner for phantom sensations simply by recourse to the body, or simply by recourse to the brain (considered as the seat of, or as identical with, the mind),[3] but that we can identify the kind of thing that would be needed to account for the phenomenon satisfactorily, and we find that this turns out to a neurological correlate of the person, as I have described this, then we will have compared the mind/brain and body accounts, on their own grounds, with a genuine third option. In other words, rather than specifying this third option in detail in advance, we need to identify just what is needed, and examine how this is best fleshed out in more general terms, building up a picture of a neurological correlate of the person from a consideration of what we need this correlate to do.

2.2 Phantom Limbs and Counterfeit Limbs

Descartes, as we have seen, believed that the medical evidence for phantom limb pain being in the brain was conclusive. But he includes within this the traditional medical explanation that the remaining nerves in the stump of the limb, which grow at the cut end into nodules, continue to generate impulses. Modern neurological accounts, by contrast, distinguish sharply between the brain and the nerve ending accounts, treating the latter as an alternative to the view that pains reside in the brain. Cutting off such nodules, or pathways within the spinal cord, can provide relief for pain for months or years, but this is the best case that can be made for a bodily source of pain, and it is far from conclusive since the sensation of the limb itself remains. Other kinds of cases, such as severe leg pain in paraplegics who have suffered a complete break of the spinal cord high in the upper body, where nerve impulses from the limb could not possibly traverse the break, indicate that such pains cannot originate in impulses in the remaining nerve endings as a general rule. The move to understanding the phenomenon of phantom limbs in terms of brain activity does not in itself resolve the issue, however, and recent work suggests that such activity is far more complex than has traditionally been thought. Information from nerve endings is processed in the somatosensory cortex, and it is quite easy to explain how, in cases of limb amputation and breaking of the spinal cord, the normal processes of neural inhibition can be over-ridden, with the result that sensory excitations arise. The trouble is that the sensory excitation that results from such neural activity alone would be relatively random. There is no explanation here of phantom limbs and their associated sensations, such as pains.

In accounting for the more specific aspects of phantom limbs, Melzack has stressed the involvement of the limbic system, as well as the somatosensory cortex. He postulates

> that the brain contains a neuromatrix, or network of neurones, that, in addition to responding to sensory stimulation, continuously generates a characteristic pattern of impulses indicating that the body is intact and unequivocally one's own...a neurosignature. If such a matrix operated in the absence of sensory inputs from the periphery of the body, it would create the impression of having a limb even after that limb has been removed (Melzack 1997, reprinted 2006, 55–56).

Such a neuromatrix involves both the limbic system, which is concerned with emotion and motivation, and those cortical regions which are 'important to recognition of the self and to the evaluation of sensory signals'. Some of the peculiarities of phantom limb behavior make a good deal more sense once we appreciate the involvement of such areas of the brain. Phantom limbs can exist in people born without a limb, for example: one patient reported in the literature, an 11-year-old girl born without hands or forearms, did arithmetic by counting on her phantom fingers. Equally puzzling if we confine our account to the somatosensory system is the fact that that an itch in a phantom limb can sometimes be relieved by physically 'scratching' in the affected area of the phantom limb (phantom limbs usually adopt a habitual position).

One part of the system that seems to be centrally involved is the parietal cortex, damage to which results in a phenomenon which I shall refer to as the counterfeit limb delusion. This delusion has been described in an engaging way in Oliver Sacks' essay 'The Man Who Fell Out of Bed' (Sacks 1985, 56–62). Sacks reports the case of a patient who was constantly falling out of bed as a result of trying to remove what he believed was a leg that had been placed in his bed, but which was in fact his own leg: as he pushed the leg out of the bed, he himself fell out also. His belief that this was not his own leg had a degree of persistence that went far beyond the sensation of sleeping on one's arm, for example, where lack of sensation in the arm can initially cause one to imagine that it does not belong to one. In the case that Sacks reports, the patient continues to deny that the leg belongs to him, even when he can see that it is attached to his body, and no matter how much others try to get him to reflect on the illogicality of his claim.

2.3   Body Image Disorders

The parallels between phantom limbs and counterfeit limbs are sufficiently striking that they are in many respects mirror images of one another. And by appreciating how they mirror one another, at the same time considering what it is that they share, namely damage to the system that checks bodily integrity, we can understand more clearly what it is that we would have to show to establish that it was, in the general case, the person, not some part of the person, that experienced sensations in general and pains in particular.

The mirroring is evident in a revealing way in the case of a phenomenon that one associates with personal response rather than brain activity or bodily activity, namely the intensity of the emotion associated with phantom limbs and counterfeit limbs. Patients with phantom limbs have a strong sense of the reality of, and ownership of, the limb, and the limbs are not experienced as at all imaginary, even in cases where a foot, for example, might be experienced as dangling several inches beneath the stump to which the original foot was connected. Indeed, prosthetic limbs can only be fitted successfully where the patient has a phantom limb, which typically is felt to fit neatly over the phantom limb. Correlatively, one of the most striking features of patients with a counterfeit limb delusion is the horror and revulsion they feel towards the limb that they believe not to be theirs. Sacks mentions the look of terror and shock on the face of the patient whose case he reports. In both the case of the acceptance of a limb one doesn't have and that of the rejection of a limb one does have, there is an emotional investment in the limb, one way or the other.

One might argue that this is exactly what we would expect if we associate both phenomena not with bodily activity but activity in the cerebrum: the emotional investment arises from the fact that both phenomena involve limbic and parietal areas of the brain. What accounts for the phenomenon is brain activity, not something more properly ascribed to the person. If this were the case, then appreciation of the complexity of what happens in the case of phantom limbs would not, contrary to what I have suggested, throw light on how we might explore the conviction that it is the person, not the person's brain, that experiences pain. The answer to this argument lies in the fact that a crucial part of what is involved – a necessary condition for both phantom limbs and counterfeit limbs – is the establishment of bodily integrity and, in particular, the existence of a faculty which acts to constantly check our bodily integrity. It is something amiss in this faculty that gives rise to the peculiar phenomenology of phantom and counterfeit limbs, a phenomenology which exhibits an emotionally-charged sense of the reality of the pathological limb.

The mirroring of the two enables us to clarify features of the phantom limb case on the question of whether the phantom limb patient can be said to be mistaken in claiming that he has a phantom limb. In the counterfeit limb case, the fact that there is indeed a mistake is manifest, and the response of doctors and nurses to try to get the patient to recognize that the offending limb is indeed his own is a natural one. By contrast, the phantom limb patient does not believe that there is a physical limb there which she cannot see. *But*, she does behave as if she has no doubt whatsoever of its reality, often unthinkingly trying to lift a cup with the phantom hand or stepping off the bed onto the phantom foot. When she behaves in this way, she is mistaken about the limb in question. Now when we say someone has made a mistake, we mean that they have done something wrong. This is why we are reluctant to say that patients with referred pains are mistaken in their natural judgments about the location of the source of pain: referral of pain is not a mistake on the part of our nervous system, but something that arises in peculiar but specifiable circumstances. Counterfeit and phantom limbs, by contrast, are not like this: they are not a case of normal functioning in peculiar circumstances. They are disorders.

They are, I have suggested, body image disorders: closely related body image disorders. They have to do with the psychological establishment of bodily integrity, 'psychological' – by contrast with 'mental' or 'cerebral' – because body images are not simply a combination of a mental/cerebral mapping and a bodily configuration but, as the emotional investment in substitution of genuine bodily parts indicates, are no longer capturable if we try to break down the person into mental and physical components, or just into physical components. The appropriate treatment for pain in phantom limbs may be bodily, such as cutting the nerves just above the nodules or in the spinal cord, or may act on the brain, such as combinations of antidepressants and narcotics, or may be more psychological in character, such as relaxation and hypnosis. But we are dealing with something that involves cognitive and affective states and necessarily, not contingently, involves not just embodiment, but embodiment in bodies of the kind that we have, that is, bodies having the kinds of limbs and organs we have, connected in the way they are.

2.4   Conclusion

In arguing for the appropriateness of a body image account of phantom limbs, my primary aim has been to set out an example of what an account of sensations that was not locked into an all-or-nothing choice between mental/cerebral and bodily accounts of pain would be like. If our aim is to capture the intuition that it is people, not their brains or bodies, who are in pain, then we need to reflect on just what is involved in pain and ask what understanding of the bearer of pain is able to accommodate what we discover about its nature, for example by examining cases such as pains in phantom limbs, which have, since the formulation of Cartesian mind/body dualism, been taken to be especially revealing about the nature, location, and sources of pain in particular and sensation in general. Yet, as I have argued, upon examination, phantom limbs, far from showing that minds/brains are the proper bearer of sensations, actually show us not just *that* people are the bearers of such sensations, but *in what way* they are bearers, helping us understand how the notion of the person might have a neurophysiological standing. Since it is in neurology and neurophysiology that much of the evidence by which to judge claims about pains in particular, and sensations more generally, is to be found, it is crucial that any option we wish to take seriously be formulable in such terms. Nor is removal of the question of the nature of pain from the realm of pure conceptual analysis a post-Quinean innovation: after all, the seminal account of the modern philosophical problem of the nature of pain, in Descartes, was motivated by considerations of the physiology of phantom limbs.

Notes

1. Descartes to Plempius for Fromondus, 3 October 1637: AT, i. 420 / Descartes 1991, 64.
2. For details of experiments and observations on phantom limb cases I have consulted: Gross and Melzack 1978; Melzack 1989; Katz and Melzack 1990; Melzack 1997; Ramachandran 1998; Ramachandran and Rogers-Ramachandran 2004.
3. Similarly for the many other accounts of the mind. I am assuming that any theory of the mind which replaced dualism and the identity theory, but which remained within the Cartesian mold, would be inadequate as an account of phantom limbs for the reasons adduced here. Other accounts, such as Rylean and Wittgensteinian behaviorism, offer different challenges, and I shall not consider these here.

References

Bennett, M. R. and Hacker, P. M. S. (2003). *Philosophical Foundations of Neuroscience.* Oxford: Blackwell.

Descartes R. (1964–1974). *Œuvres*, eds C. Adam & P. Tannery, 11 vols. Paris: Vrin (cited as AT followed by volume and page number).

Descartes, R. (1991). *The Philosophical Writings of Descartes,* Volume III, trans. J. Cottingham *et al.* Cambridge: Cambridge University Press. (CSMK).

Gross, Y. and Melzack, R. (1978). Body Image: Dissociation of Real and Perceived Limbs by Pressure-Cuff Ischemia. *Experimental Neurology,* 61(3), 680–688.

Katz, J. and Melzack, R. (1990). Pain "Memories" in Phantom Limbs: Review and Clinical Observations. *Pain,* 43(3), 319–336.

Melzack, R. (1989). Phantom Limbs, the Self and Brain: The D. O. Hebb Memorial Lecture. *Canadian Psychology,* 30(1), 1–16.

Melzack, R. (1997). Phantom Limbs. *Scientific American* 266, 84–89; reprinted in *Scientific American* Special Edition "Secrets of the Senses", 16(3), (2006), 52–59.

Ramachandran, V. S. (1998). *Phantoms in the Brain: Human Nature and the Architecture of the Mind.* London: Fourth Estate.

Ramachandran, V. S. and Rogers-Ramachandran, D. (2004). Phantom Limbs. In Richard L. Gregory, ed., *The Oxford Companion to the Mind*, 2nd edn. Oxford: Oxford University Press. Also on Oxford Reference Online: http://www.oxfordreference.com.

Sacks, O. (1985). *The Man Who Mistook His Wife for a Hat.* London: Duckworth.

3
Is the Next Frontier in Neuroscience a 'Decade of the Mind'?[1]

*Jacqueline A. Sullivan**

3.1 Introduction

In 2007, ten world-renowned neuroscientists, including James Albus, George Bekey, John Holland, Nancy Kanwisher, Jeffrey Krichmar, Mortimer Mishkin, Dharmendra Modha, Marcus Raichle, Gordon Shepherd, and Giulio Tononi advocated in a letter published in *Science* for "a major national research initiative called 'A Decade of the Mind'" (Albus et al. 2007, 1321). Their contention was that, despite the successes of the *Decade of the Brain*, "a fundamental understanding of how the brain gives rise to the mind [was] still lacking" (ibid.). They identified four areas of research to be the focus of this new decade including: (1) healing mental disorders; (2) understanding "aspects of mind believed to be uniquely human" including "the notion of self, rational thought processes, theory of mind, language and higher order consciousness"; (3) "enriching the mind through education"; and (4) "modeling the mind by means of computational models and artificial intelligence" (ibid.). The proposed decade was to be "transdisciplinary and multi-agency in its approach," incorporating insights from neuroscience, medicine, cognitive neuroscience, psychology, computer science, engineering, mathematics, robotics, systems biology, cultural anthropology and social science (ibid.).

Six years have passed since the publication of the letter, and only one direct response to it has appeared in the scientific literature. However, that response, provided by the German psychiatrist Manfred Spitzer, did not take up the most interesting issues raised by the proposal. Spitzer interpreted the authors as merely suggesting that, given that research during the decade of the brain had been directed primarily at under-standing the mechanisms of "perception and motor control," it was time for "investigators in systems neuroscience" to "turn their attention and a powerful arsenal of methods towards what traditionally were regarded as 'mind-functions'" (Spitzer 2008). In other words, Spitzer interpreted the authors of the proposal as merely advocating for the broader applica-tion of those investigative strategies prevalent in cognitive and systems neuroscience during the 1990s to more complex phenomena such as theory of mind, the self and higher-order consciousness. While it may just be that Albus and colleagues' call for a decade of the mind was supposed to involve the extension of imaging technologies to study higher-order mental functions and dysfunctions, that hardly seems like the "paradigm-shifting progress" and interdisciplinary call to arms that the proposal was intended to instigate. Thus, the only response to the proposal in the review literature downplayed what was truly interesting about it, namely, that a group of world-renowned neuroscientists had acknowledged that a major change in how neuroscientists study the mind-brain relationship was needed.

The aim of this chapter is to remedy this oversight by addressing what I regard as a set of interesting questions that the proposal for the decade of the mind prompts – questions that I believe, when answered, enable a more cogent case for a decade of the mind and provide insight into how to implement the proposal in neuroscientific practice.[2] First, what is "the mind" that is supposed to be the target of this new decade? Second, what was missing during the decade of the brain that would prompt the need for a separate decade of the mind? Third, what should this new decade look like?

3.2   What is 'the mind'?

One fundamental aspect of being human is that we learn to adopt what Daniel Dennett (1987) has dubbed "the intentional stance." Specifically, we come to believe that human beings as well as some non-human animals have some special quality – a *mind, consciousness, awareness* – that other kinds of things – rocks, stars, and trees – lack. We describe ourselves as having beliefs, desires, feelings, and intentions. We ascribe similar internal states to other human beings and some non-human animals. We appeal to these states to explain our own and others' behav-iors. This conceptual-explanatory framework, which has been elevated to the status of a theory by cognitive psychologists and philosophers (for example Churchland 1981), plays a fundamental role in introspection and in our development of a concept of self and a personal identity. It also has and continues to serve as a principal basis for philosophical thinking about the nature of consciousness, subjective experience, mental causation, cognition and knowledge, ethics, the nature of the self, personal identity, and mental disorders (e.g., delusions as false beliefs). Insofar as components of this framework may vary across cultures, races, and religions, and such differences may be put forward to explain social, economic, political, historical, and cultural phenomena, it plays a fundamental role in the humanities and in the social sciences,[3] including those areas of science the proponents of the Decade of the Mind (DoM) would like to be involved in this new initiative.

The DoM proposal may be misunderstood as a call for a revival of folk psychology as a conceptual-explanatory framework for at least two reasons. First, the authors define what they mean by "mind" ostensively rather than discursively, and the terms put forward by them including "mind," "higher-order consciousness," "the notion of self," "rational thought processes," and "mental disorders" overlap with the mixed ontology of entities and activities that function as causes of behavioral phenomena in the folk-psychological worldview. A second and related reason why the proposal appears sympathetic to folk-psychological theory is that the initiative is supposed to be "transdisciplinary" – it is intended to instigate interdisciplinary interactions among neuroscientists and practitioners coming from more disparate research areas in which "talk about belief is ubiquitous" (Dennett 1987, 13). The success of the decade as conceived by the DoM authors purportedly requires crosstalk among investigators coming from areas of science that take folk psychology seriously and areas of science that do not. It would be disingenuous for investigators to attempt to include practitioners who regularly appeal to folk-psychological explanations in the DoM initiative while at the same time devaluing this approach. Third, one aim of the initiative is "educating the general public on legal and ethical issues involving the brain and the mind" (Albus et al. 2007, 1321). Effectively communicating scientific results to the general public whose primary understanding of the mind, mental disorders, consciousness, and the self is rooted in folk psychology seems to require some attempt to take that conceptual-explanatory framework seriously.[4]

Other facts, however, may be taken to indicate that the proponents of the decade of the mind are not interested in the conceptual explanatory framework of ordinary folk.[5] First, what we find when we look across the contemporary neurosciences are multiple distinct areas of science ranging from molecular genetics to behavioral neuroscience and "a plurality of incompletely articulated and partially contradictory, partially supplementary theories and models" (Wimsatt 2007, 180) that are all purportedly directed at understanding the neural mechanisms of consciousness and cognition as well as the brain dysfunctions that give rise to mental disorders. Folk psychology as a conceptual-explanatory framework is notably absent from the diverse array of explanations that the contemporary neurosciences yield for these phenomena. This makes sense because a primary aim of the biological and physical sciences is to move beyond folk understandings and explanations of phenomena to discover their *physical* mechanisms.[6] The mental states of the folk do not fit into the world in any interesting way; they are not parts of the ontological hierarchy that runs the gamut from molecules to behavior. Secondly, given that the investigators who authored the DoM proposal work (or did work) in a variety of areas of the cognitive and neuro-sciences including: artificial intelligence (e.g., Modha), engineering (e.g., Albus, Holland), computer science and robotics (e.g., Bekey, Holland), cognitive science (e.g., Krichmar) cognitive neuroscience (e.g., Kanwisher, Mishkin, Tononi) neuropsychology (e.g., Mishkin), psychiatry (e.g., Tononi), neurobiology (e.g., Shepherd), and neurology (e.g., Raichle), it is reasonable to conclude that they are endorsing the conceptual-explanatory framework of cognitive science and with it the *cognitive notion of mind.*

According to the cognitive notion, mind is the total set of an organism's cognitive states and processes that are causally responsible for, but not identical to, its overt behavior. The conceptual-explanatory framework of cognitive science contains a basic set of assumptions that practitioners coming from those diverse fields represented in the DoM proposal share in common. These assumptions include most basically that (1) human beings have specific kinds of cognitive capacities, (2) these capacities involve "representational structures and processes," and insofar as (3) these representational structures and processes carry information about what they represent (4) thinking of the mind as an information-processing device is a fruitful analogy (von Eckhardt 1993, 1). This set of assumptions is clearly not without its problems, because the "representational structures" of cognitive science that "carry information" are purportedly *mental* states. What relationship mental states bear to neural states and the question of how neurons can assume the representational capacities normally ascribed to them remain subjects of scientific and philosophical debate (for example Bechtel 2008; DeCharms and Zador 2000). So, proposing to revive the cognitive notion of mind in neuroscience has the potential to revitalize a host of thorny problems that the decade of the brain did not resolve, but left behind. However, a persuasive case can be made for reviving the cognitive notion of mind in contemporary neuroscience. In order to make this case, I regard it as first relevant to consider how philosophers have conceived of the relationship between folk psychological and cognitive psychological explanations of cognitive phenomena, so I can be very clear about precisely what aspects of the conceptual-explanatory framework of cognitive science that go hand-in-hand with the cognitive notion of mind need to be revived and where.

Some philosophers, most notably Jerry Fodor (1975), equate the mental states of the folk with the so-called "propositional attitudes," statements of the form "S _____ that P." where S is the subject (e.g., Wayne, Mandy, Tim), P is a proposition (e.g., "All mental states are brain states.") and "_____" may be filled in with any one of a number of verbs expressing a disposition towards that proposition (e.g., believing, hoping). These same philosophers have implied that insofar as cognitive psychological explanations make reference to mental states, and mental states are nothing over and above propositional attitudes, cognitive psychological explanations and folk psychological explanations are similar.

Robert Cummins (1983) most notably has responded that such claims are based on a misunderstanding of the nature of psychological explanation, which has as its aim to explain the cognitive capacities of organisms by *functional analysis*. Explanation by functional analysis, according to Cummins, "does not traffic in explicit propositional attitudes" and "does not involve intentional characterization" (Cummins 1983, 82). Rather, the aim of functional analysis is to ascribe cognitive capacities to organisms and identify the sub-capacities that enable the realization of those capacities. To take a simple example, organisms can remember things; they have the cognitive capacity of memory. However, organisms remember different kinds of things – not only how to perform certain procedures but also that certain declarable facts obtain. In other words, memory may be broken down into at least two sub-capacities, which are not propositional attitudes and are not necessarily characterized in terms of them. Ideally, the concepts of "procedural memory" and "declarative memory" designate sub-capacities that can be ascribed to divisible systems in the brains of organisms that exhibit the cognitive capacity. Neuroscience has in fact shown this to be the case, with declarative memory being subserved by structures in the medial temporal lobe (e.g., the perirhinal and parahippocampal cortices, the entorhinal cortex, and the hippocampus) and procedural memory being subserved by the striatum, basal ganglia, and cerebellum. Cognitive psychological explanations thus differ from folk psychological explanations because they ultimately require investigators to take the *physical stance* towards their objects of inquiry.

Although Cummins is correct about the basic aims of explanation in cognitive psychology, I think the claim that cognitive psychological explanation "does not involve intentional characterizations" is false. For it is based on the false assumption that taking the intentional stance towards an organism and ascribing propositional attitudes to it is the same thing. But why equate them? Even the common person deploys a mixed ontology of mental states (anger, happiness, attention, memories) and mental processes (believing, desiring, wanting, remembering, attending) in order to explain and predict behavior. Thus, it is incorrect to claim that when ordinary folk adopt the intentional stance they are "trafficking" exclusively "in propositional attitude ascriptions." Propositional attitude theory offers just one conceptual-explanatory framework for explaining the behavior of the folk and understanding what they do when they explain their own and others' behaviors. However, when cognitive scientists "talk...about human cognitive activities" they may be interpreted simply as "speak[ing] about mental representations and...posit[ing] a level of analysis wholly separate from the biological or neurological" level (Gardner 1985, 6). Propositional attitude theory does not capture what they do, but claiming that they take the intentional stance towards their objects of inquiry, namely human cognizers, does. If we grant this, then, contrary to what Cummins claims, it *is* possible that cognitive psychologists in particular and cognitive scientists more generally take *the intentional stance* towards those organisms whose capacities they are interested in explaining. Let's call this a modified version of the intentional stance insofar as the achievement of their explanatory aims requires the positing of abstract mental states and mental processes. Perhaps sometimes this includes propositional attitude ascriptions, but that is not a requirement. This brings us to the question of *when* in the process of doing science cognitive psychologists take the intentional stance, given that, as Cummins rightly claims, explanations by functional analysis, in the final analysis, do not involve intentional characterizations – a point I, and recent proponents of mechanistic explanation (e.g., Bechtel 2008; Craver 2007; Piccinini and Craver 2011) think he is correct about.

One answer to this question that is implicit in Cummins (1983) and explicit in Fodor (1968) is the idea that the intentional stance is involved *only in the earliest stages of* psychological explanation. Cummins (1983) refers to two kinds of functional analysis: *dispositional analysis*, which involves breaking down a capacity into abstract sub-capacities that are ascribed at the level of the whole intact organism and *componential analysis*, which requires that these sub-capacities be ascribed to the internal physical components (e.g., brain structures, synapses, neurons) of the organism that exhibits them. Whereas dispositional analysis requires that cognitive psychologists take the intentional stance, componential analysis does not.[7] These two types of analysis correspond directly to what Fodor (1968) identifies as two phases of psychological explanation. In the first phase, "the psychologist is seeking functional characterizations of psychological constructs" and "the criteria employed for individuating such constructs are based primarily on hypotheses about the role they play in the etiology of behavior" (Fodor 1968, 107–108).[8] The constructs employed by psychologists at this stage of explanation are hypothetical and abstract. Determining whether these constructs correspond to actual divisions in the brain is the second phase of psychological explanation. During this phase the psychologist aims to identify "those biochemical systems that do, in fact, exhibit the functional characteristics" of interest and she "'looks inside' to see whether or not the nervous system does in fact contain parts capable of performing the alleged functions" (ibid., 109). Given Cummins' and Fodor's claims, then, it is reasonable to conclude that whereas the first stage of psychological explanation requires that investigators take the intentional stance towards their experimental subjects, the second phase requires them to take the physical stance and assume that an organism's behavior has internal physical (e.g., neural, biochemical) causes.[9]

I want to suggest, however, an even earlier and more important role for the intentional stance in cognitive psychology in particular and in the neurosciences of cognition more generally. As Fodor claims, in the early stages of explanation psychologists posit "hypothetical constructs." In cognitive psychology in particular and cognitive science more generally, a construct is a postulated capacity or attribute of an organism that may also be a target of psychological explanation. Intelligence, working memory, innateness, and spatial memory are all examples of constructs. A construct typically originates, "with a vague concept which we associate with certain observations" (Cronbach and Meehl 1955, 286), and this vague concept serves as basis not only for theory building in psychology (as well as other social sciences) but also for designing experiments to study cognitive capacities of interest, and in the case of the neurosciences of cognition, localizing them in the brain and determining the neural, synaptic, cellular, and molecular mechanisms that give rise to them. While the origin of such vague concepts may be the conceptual-explanatory framework of folk psychology, investigators also rely on definitions already available in the scientific literature – particularly definitions that are commonly deployed in that same field of research.

When a cognitive psychologist goes into the laboratory, she will aim to design an experimental paradigm or cognitive task in order to investigate the cognitive capacity of interest. An experimental paradigm is roughly "a standard set of procedures for producing, measuring and detecting" a cognitive capacity "in the laboratory" that "specifies how to produce" that capacity, "identifies the response variables to be measured during pre-training, training, and post-training/testing" and includes instructions on "how to measure [those response variables] using equipment that is designed for this purpose" (Sullivan 2010b, 266). It also specifies how to detect the cognitive capacity "when it occurs, by identifying what the comparative measurements of the selected response variables have to equal in order to ascribe" that capacity to a subject (ibid.). Ideally, the investigator will aim to design an experimental paradigm capable of reliably individuating the cognitive capacity of interest that the construct purportedly designates. Individuating cognitive capacities requires the development of experimental paradigms or cognitive tasks that are *reliable* for this purpose. Ensuring the reliability of an experimental paradigm requires, I am claiming, that an investigator adopt the intentional stance towards her subjects. This is supported by the fact that when cognitive neuroscientists design experimental paradigms they regard it as important to consider what may ultimately go on inside the head of a hypothetical subject when that subject is trained and tested in that paradigm. The vast majority of cognitive neuroscientists, particularly those with training in cognitive psychology, assume that there is such a thing as mental function, that it "is composed of distinguishable fundamental processes" and that "these processes can be selectively engaged by properly designed experimental task manipulations" (Carter et al. 2009, 169). In light of this assumption, some cognitive neuroscientists engage in rigorous "theoretically guided task analys[es]" intended to provide "a clear specification of the processes thought to be engaged by an experimental task and how these processes will be influenced by the variables to be manipulated in the experiment" (ibid., 169).[10] Designing tasks that successfully individuate a cognitive function and allow for its localization in the brain requires a significant amount of ingenuity. If an investigator fails to consider the possible mental states of her hypothetical subject or she neglects to itemize other hypothetical or actual processes that may be involved in the execution of the task (e.g., attention, working memory), then the task that she designs will likely be unreliable for individuating the cognitive capacity of interest to her. The task will lack *construct validity* if there is a discrepancy between the hypothetical construct that the task was intended to measure and those cognitive processes actually measured.

Investigators across the contemporary cognitive and neurosciences have substantial freedom to produce, detect, and measure cognitive functions using the experimental paradigm or task that they take to be most reliable for achieving their investigative aims. Not all investigators will agree that a particular experimental task or paradigm is subject to one exclusive task analysis or that it measures a discrete cognitive function. In fact, disagreements about the potential functions that play a role in the execution of a given task may prompt revisions to that task and/ or the development of new tasks. For example, the Stroop task,[11] which was for a long time widely thought to individuate the cognitive capacity of *selective attention*, has also been described as measuring *response inhibition* and *context processing*. Precisely what cognitive function it individuates remains a subject of debate (Perlstein et al. 1998; Cohen et al. 1999; Barch et al. 2004). Many tasks have prompted similar debates. However, it is well recognized that developing tasks that successfully individuate cognitive capacities is an iterative, trial-and-error process. What I am claiming is that in cognitive psychology and cognitive neuroscience it is an iterative process that intimately involves the intentional stance.

Insofar as cognitive neuroscientists adopt the intentional stance in the experimental context, we can be rest assured that the cognitive notion of mind is alive and well in contemporary cognitive neuroscience. If this is correct, then the proposal for a decade of the mind in neuroscience must be directed at some other target. For a variety of reasons, I think the primary target is *cognitive neurobiology*. First, the authors of the proposal stress that the aim of this new decade is to move past the achievements of the decade of the brain, which "focused on neuroscience and clinical applications" (Albus et al., 2007, 1321). It is reasonable to suspect that they have low-level neuroscience in mind here, which focused primarily on the development of treatments for mental disorders and the identification of the cellular and molecular mechanisms of learning and synaptic plasticity during the decade of the brain. As John Bickle claims, the real "'revolution'" in neuroscience during that decade occurred in "cellular physiology and molecular biology" (2003, 2) rather than in cognitive neuroscience. Bickle (2003, 2006) also has correctly pointed out that investigators in cognitive neurobiology do not make reference to mental functions or to the mind. Rather, they take the physical stance towards their experimental subjects, they operationally define cognitive capacities in terms of observable changes in behavior and then look directly into the brain to determine the cellular and molecular activity implicated in the production of those changes in behavior (see also Sullivan 2009, 2010a, 2010b; Sweatt 2009). Third, the authors of the proposal stress that this new decade should be "transdisciplinary and multi-agency" in its approach. In other words, it requires scientists in all the areas that are to be involved in the initiative to take seriously the cognitive notion of mind. This makes sense because determining how the brain gives rise to the mind will be a non-starter if investigators in the business of localizing cognitive functions in the brain (i.e., cognitive neuroscientists) cannot effectively communicate with investigators who are discovering their cellular and molecular mechanisms (i.e., cognitive neurobiologists).

The call for "paradigm-shifting" progress in neuroscience conjures up Kuhn's (1962) notion of a paradigm and the problem of incommensurability: investigators working in radically opposed scientific traditions that are associated with different foundational assumptions and investigative approaches have trouble communicating to the extent that it sometimes seems as if they are living in different worlds. In fact, cognitive neuroscience and cognitive neurobiology do indeed emanate from separate and radically opposed historical traditions, and appreciating this history enables us not only to recognize potential obstacles to this new decade but also to appreciate that overcoming such obstacles requires reviving the mind in the experimental context in cognitive neurobiology.

3.3   Why revive the cognitive notion of mind?

Cognitive science as we know it today began to emerge in the second half of the 20th century when a group of scientists representing a diverse array of disciplines including mathematics, computer science, neurophysiology, and psychology began to search for methods to understand the mind, the brain, and behavior that were alternatives to the then-dominant experimental psychology of behavior (Gardner 1985). The "behaviorists," including J.B. Watson, E.L. Thorndike, Edwin Guthrie, Clark Hull, and later B.F. Skinner, aimed to develop a scientific psychology that was on a par with the physical sciences. The behaviorists took the achievement of this aim to require the rejection of 19th-century introspective psychology,[12] its method of introspection and "its subject matter *consciousness*" (Watson 1924[1970], 2). In its place they put forward a science that was supposed to rely exclusively on publically verifiable methods, explaining behavior solely by appeal to stimuli and responses. They supposed that the experimental learning paradigms of classical and operant conditioning – two standard sets of procedures for producing, measuring, and detecting forms of associative learning in the laboratory – could be used to investigate the causes of an organism's behavior without investigators needing to concern themselves with what was going on inside those organisms' heads. Another way of putting it is that they rejected "the intentional stance" in psychology; it played no role in the process of designing and implementing experimental learning paradigms or in explaining the data.

Behaviorism and the experimental paradigms of classical and operant conditioning came under attack for a number of reasons. First, it was difficult for experimental psychologists to reliably isolate learning conditions that were favorable exclusively to stimulus-stimulus and stimulus-response type explanations because it was often difficult to identify precisely what the independent variables were. For example, as J.J.C. Gibson (1960) and Charles Taylor (1964) claim, it was difficult to assess to which stimuli or which aspects of those stimuli organisms trained in classical and operant conditioning paradigms were actually responding. Another limitation was that the experimental paradigms were neither ecologically nor externally valid. The highly artificial conditions of the laboratory, which included raising organisms in impoverished environments and depriving them of an experiential history, was regarded by critics as an obstacle to generalizing from learning in the laboratory to learning in the world (see e.g. Hinde 1973, Lorenz 1965). A more severe problem was that learning in classical and operant conditioning paradigms required the repetition and contiguity of stimuli or stimuli and responses. However, as the psychologist Karl Lashley argued in his lecture at the Hixon Symposium, more complex forms of learning (e.g., Kohler 1947) did not appear to conform to this associative learning model. As the historian Howard Gardner claims, "so long as behaviorism held sway [...] questions about the nature of human language, planning, problem solving, imagination, and the like could only be approached stealthily and with difficulty, if they were tolerated at all" (Gardner 1985, 12).

Physiologists trained in the behaviorist tradition, however, were not moved by Lashley's worries about the limitations of associative learning theory for explaining complex cognitive capacities (e.g., learning a language or learning how to play a sport or an instrument). Some also doubted the reliability of Lashley's ablation experiments, which he used to refute the widely accepted idea that memory traces were stored in the brain in those neurons activated during training in associative learning paradigms. In a seminal review paper in 1968, Eric Kandel and W. Alden Spencer sought to discredit Lashley in claiming that his "experimental techniques and his conclusions had been seriously questioned" (1968, 67). Feeling that they had successfully turned back the only challenge to the idea that "the synapse" plays "a crucial role in information storage" (1968, 66), they emphasized the importance of the repetition and contiguous presentation of stimuli to forge new synaptic connections in the brain.

Kandel and Spencer essentially echoed the ideas of the neobehaviorist Donald Hebb. In 1949, Hebb, borrowing insights from behaviorism, gestalt psychology and physiology, suggested that learning and memory are achieved by physiological changes in brain synapses. Hebb claimed that when two cells, A and B, which communicate under normal conditions, undergo a period of repeated and concurrent activation, as may happen during classical or operant conditioning, the result will be a strengthening of the connection between the two cells. According to Hebb, this strengthening is reflected in a subsequent change in the way the one neuron excites the other (Hebb 1949). The crux of this postulate, often referred to as "Hebb's rule," is that each associative learning event is accompanied by the brief associated activation of two neurons that comprise a synapse, which together, effectively store information in the form of a physiological change at that synapse.

Although Hebb's postulate was attractive, at the time of its introduction, no plausible candidates for a neural mechanism of associative learning that satisfied his described conditions had been located in the mammalian brain. However, in 1966, in the context of investigating the physiology of the dentate gyrus in the hippocampus of the adult anesthetized rabbit, Terge Lømo observed an artificially induced physiological equivalent of a strengthening in synaptic efficacy (Lømo 2003). This discovery of a "long-lasting potentiation" in area CA1 of the rabbit hippocampus *in vivo* led to Lømo's famous publication with Tim Bliss in 1973 in which they described the phenomenon of long-term potentiation (LTP), which instantiated all of the features of the mechanism of associative learning that Hebb (1949) had described (see Craver 2003).

Kandel's seminal research on the cellular and molecular mechanisms of learning and memory in the sea mollusk *Aplysia Californica* in combination with Tim Bliss and Terge Lømo's (1973) discovery of long-term potentiation (LTP) in area CA1 of the rabbit hippocampus *in vivo* form the cornerstones of modern cognitive neurobiology. Although the paradigms of classical and operant conditioning and associative learning theory were rejected by those attendees of the Hixon symposium who founded modern cognitive science, they found a good home in cognitive neurobiology where versions of the two learning paradigms are widely used to this day in conjunction with electrophysiology experiments that are used to induce LTP.

Paradigms intended to investigate more complex cognitive functions also have been introduced into the cognitive neurobiological literature. Social recognition memory paradigms (Bickle 2006; Sullivan 2009) and the Morris water maze (e.g., Craver and Darden 2001; Craver 2007; Sullivan 2010a, 2010b) are some widely celebrated examples. However, their introduction has led to certain kinds of problems in part because of the failure on the part of cognitive neurobiologists to take the intentional stance. Specifically, investigators in cognitive neurobiology are not interested in "what" cognitive capacity is operative when an organism is trained in an experimental learning paradigm. They are satisfied just so long as they can use those paradigms to produce robust behavioral effects in which they can pharmacologically or genetically intervene. When they train organisms in experimental learning paradigms, they do not assume that these organisms have minds; they do not take the intentional stance, but the physical stance.

One representative example is the hidden condition of the Morris water maze. The water maze is an open field maze consisting of a large circular pool filled with opaque water. The pool is placed in a room containing a discrete set of fixed distal visual cues. In the hidden condition of the water maze, a silvery-white platform is placed just beneath the water's surface so as to be undetectable to a rodent placed in the pool. During training in the hidden condition, the location of the platform remains fixed across trials and the placement of the rat in the pool varies randomly with respect to the four cardinal positions (i.e., N, S, E, W). When a rat is placed into the pool, it will attempt escape, and thus swim about the pool. On each training trial, the swim path of the animal in the maze, the length and direction of the angle of that path, and the time it takes it to find the platform ("escape latency") are measured. A significant decrease in the amount of time it takes the animal to find the hidden platform across training trials is taken to indicate that the rat has learned the location of hidden platform solely on the basis of the distal room cues. Morris originally referred to this set of behavioral effects observed in the water maze as "place learning," which was intended to capture the idea that the rats learned the place of the hidden platform solely on the basis of the distal room cues, rather than by stimulus-stimulus or stimulus-response associations.

An historical analysis of the Morris water maze, however, reveals that over a 30-year time span, across the experimental and review literature in cognitive neurobiology, the term used to designate the phenomenon under study in the hidden condition of the maze oscillated. Candidate terms included place learning, place navigation, spatial learning, spatial memory, spatial navigation, water maze navigation, and water maze performance (Sullivan 2010b). Such oscillations suggest not only that investigators were unclear what cognitive function was under study in the water maze, but also that over a 30-year time span, only slight efforts were directed at achieving clarity. This makes sense given that cognitive neurobiologists are not concerned with "what" rodents trained in the water maze learn or what cognitive functions are involved in the production of the behavioral effects. After all, they do not assume organisms have minds, nor do they take the intentional stance towards their experimental subjects. However, this lack of concern is an impediment to explanatory progress because to discover the mechanisms of a cognitive function, it is necessary to know what the function is (see for example D'Hooge and De Deyn 2001).

One of the reasons the water maze is such an interesting experimental paradigm is that it teaches us lessons about the challenges of the scientific study of cognitive functions. The first lesson is that when an experimental paradigm is designed, establishing its reliability for individuating a discrete cognitive function requires a consideration of "what" an organism trained in the paradigm is learning. This suggests that investigators must appeal to a cognitive understanding of the mind, take seriously the potential mental states and information processes of the whole intact organism and engage in a thorough task analysis, much like cognitive neuroscientists do. Secondly, when an experimental paradigm is used to reliably produce a discrete set of behavioral effects, investigators often assume that they have individuated the cognitive function of interest, and the search for the systems, synaptic, cellular, and molecular mechanisms productive of those behavioral effects then begins. However, cognitive functions are not identical to behavioral effects that result from training an organism in an experimental paradigm. The causes of the changes in behavior likely include many more changes in internal states and processes than are captured by the term designating the cognitive function under study in the paradigm (see for example Taylor 1964). This is why taking the intentional stance is not only important in the context of designing experimental learning paradigms, but it is also fundamental when such paradigms are being implemented in the laboratory and the results of these experiments are being interpreted.

Currently, the possible limitations of all the experimental paradigms used in cognitive neurobiology are likely being missed precisely because cognitive neurobiologists fail to take an organism's mental states seriously once they have identified an experimental paradigm that seems to produce robust behavioral effects. However, the problems that arise from failing to raise questions about "what" an organism is learning, remembering, or doing, and what representational processes are involved, are an impediment to individuating discrete explanatory targets and identifying their mechanisms. To eliminate such problems and thus answer the question of where we need to revive the mind in contemporary neuroscience, I think the answer is that the cognitive notion of mind and the intentional stance ought to play a fundamental role in the experimental context in cognitive neurobiology when investigative strategies are being designed and/or implemented.

Ironically, the proposal for a decade of the mind in contemporary neuroscience is roughly similar to the original call to revive the mind made by the diverse array of scientists who attended the Hixon Symposium in 1948 where the seeds of contemporary cognitive science were sown. The scientific backgrounds of those members of that original group are also similar to the backgrounds of those who gathered at George Mason University in 2007 to discuss the need for a new decade of the mind. Thus, in answer to the question of why a world-renowned group of cognitive scientists would call for a new decade in contemporary neuroscience, I think we need only look to the reasons put forward by the attendees of the Hixon symposium who recognized that the only way to overcome the limitations and problems with the then-current conceptual-explanatory framework for understanding learning and memory was to revive the mind in psychology and bring back the intentional stance.[13]

3.4   What should a decade of the mind look like?

Proponents of decade of the mind argue that this new decade should be "broad in scope" and "transdisciplinary in nature." However, they offer no positive proposals with respect to what form such interdisciplinary interactions should take. Given the insights revealed from my analysis of the Morris water maze, I have indicated that one appropriate venue for collaboration is the experimental context. More specifically, I am claiming that to increase the reliability of experimental paradigms for individuating cognitive functions, practitioners from a variety of different areas of the mind-brain sciences, including but not limited to: cognitive psychologists, cognitive neuroscientists, experts in animal behavior, computational scientists, and molecular and cellular cognitive neurobiologists, should combine forces to develop and implement experimental paradigms in the laboratory. Furthermore, extensive dialogue across research teams and disciplines using the same experimental paradigm should be on-going. Such interdisciplinarity makes good sense for several reasons. First, in any area of science that uses whole intact organisms, one must be privy to the fact that a variety of different processes – molecular, cellular, synaptic, network, systems, representational, informational, and behavioral – co-occur simultaneously. Second, different investigators, given different areas of expertise, have different explanatory interests, and they face different obstacles in developing and implementing experiments that work for their distinct explanatory purposes. These explanatory interests and obstacles require a forum within which solutions may be located and the impact of such solutions on the phenomena under study (i.e., cognitive capacities) may be considered.

It makes sense for such "perspectival pluralism" (Giere 2010; Wimsatt 2007) to be implemented in the context of experimentation rather than the context of explanation in neuroscience for several reasons.[14] To date, the various areas of science that comprise the contemporary neurosciences have yielded a plurality of piecemeal explanations of cognitive phenomena that do not fit together in any interesting way – in part because there is no standardization of the use of concepts designating constructs in contemporary neuroscience[15] and investigators working at all levels of analysis are free to develop cognitive tasks working with whatever assumptions about their experimental subjects they regard as germane to their research.[16] Yet, what investigators seem to want are coherent multi-level mechanistic explanations of cognitive phenomena (Bechtel 2008; Craver 2007). Although philosophers of neuroscience have suggested that such explanations are on the horizon and that "explanatory unification will be achieved through the integration of findings from different areas of neuroscience and psychology into description[s] of multilevel mechanisms" (Piccinini and Craver 2011, 285), they never specify how one area of science that strives to ensure that its investigative strategies individuate discrete cognitive functions will be able to readily integrate its findings with an area of science that does not.

However, we can point to examples in the neuroscientific literature that support the claim that maintaining the current status quo in neuroscience is insufficient and that "paradigm-shifting progress" is necessary if we want to solve important problems like eradicating mental illness.

One representative example is an interdisciplinary research initiative that has evolved during the past 6 years with the aim of developing effective "pro-cognitive" agents to eliminate cognitive deficits in schizophrenia. The teams of investigators involved in the *Cognitive Neuroscience Treatment Research to Improve Cognition in Schizophrenia (CNTRICS)* Initiative regard schizophrenia as involving a core set of cognitive dysfunctions. A primary aim of the initiative is to develop experimental paradigms/cognitive tasks capable of individuating those cognitive functions that are disrupted in schizophrenia. Given that the ultimate aim of these research treatments, which first have to be tested in animal models, the initiative has required cognitive neuroscientists to interact with cognitive neurobiologists, experts in animal behavior, clinical pharmacologists, and members of industry. Their first aim was to develop cognitive tasks that could be used to identify which cognitive functions are disrupted in schizophrenia. However, in developing these paradigms they also had to concern themselves with what kinds of cognitive functions it is possible to study in animal models. This required matching tasks and engaging in task analysis across species. The initiative is still in progress, with different task forces directed at solving different kinds of practical problems. However, what is important for my purposes is that the investigators who are involved in the initiative are interested in bringing about "paradigm-shifting" progress in neuroscience, they believe that this requires interdisciplinary dialogue in the contexts of designing and implementing experimental learning paradigms, and they also believe that on-going interdisciplinary dialogues are fundamental to the success of the initiative (see Sullivan forthcoming).

On a final note, some of the aspects of mind that the DoM authors identify, including "the notion of self" and "higher-order consciousness," are not as readily construed as cognitive capacities, and little work has been undertaken to design investigative strategies for individuating them. This is where insights and perspectives from cognitive psychology, clinical medicine, cultural anthropology, and even philosophy may be helpful. To take one interesting example, Sadhvi Bahtra and colleagues (2013) have developed a semi-structured interview to operationalize the self, so that it may be elicited in terms of behavioral responses, much like cognitive capacities are. One reason for developing this questionnaire is to improve the treatment of patients who suffer from memory disorders like Alzheimer's disease (AD). Even despite the fact that "Alzheimer's disease is often characterized as leading to 'a loss of self' [...] people with dementia continue to refer to themselves with 'I' and often do not recognize the cognitive deficits ascribed to them"

(Bahtra, Geldmacher, and Sullivan 2013). Developing an investigative tool that allows for the qualitative assessment of the self may provide clinicians with clues on how to improve the conditions of life of persons who have AD. Of course, the benefits of developing such investigative tools may not stop there. If successful, such self-assessment questionnaires may one day evolve into investigative tools that may be effectively combined with functional imaging technologies, with the ultimate aim being to determine what brain structures subserve the self. However, the self is a construct, and developing a tool/procedure/task that effectively individuates it will be a challenging iterative process. Fundamental to the success of this process will be a plurality of different perspectives involved in designing and implementing investigative strategies for studying it as well as on-going interdisciplinary discussions about how such strategies succeed and/or fail.

3.5 Conclusion

The proposal for a decade of the mind in 2007 generated little attention among practicing neuroscientists and practitioners working in areas of science targeted as fundamental players in this new decade. Implicit in the questions I have sought to answer in this chapter is a criticism of the proposal for failing to answer a set of questions that are relevant for generating interest in such a decade and making a case as to why such a decade is essential to the future success of neuroscience with respect to understanding how the brain gives rise to the mind.

The first problem with the proposal is that the proponents of the initiative did not explain what they meant by mind. I have here sought to identify what notion of mind they regard as relevant. My claim is that they want to revive the cognitive notion of mind in contemporary neuroscience and with it the intentional stance.

The proposal also lacked a clear motivation. I have provided one such motivation in demonstrating that the cognitive notion of mind and the intentional stance offer two valuable perspectives that should be operative when investigators are designing and implementing tasks that are intended to successfully delineate cognitive functions. Furthermore, when they fail to be operative in the experimental context, the explanatory aims of neuroscience cannot be realized.

A third problem with the proposal is that the authors never specified what this new decade was supposed to look like. They failed to provide a set of guidelines for how to get such an interdisciplinary initiative off the ground. I have suggested that one way in which we can increase the probability of investigators from multiple disciplines coming together to solve common explanatory problems of interest is to sanction perspectival pluralism in experimental contexts across the neurosciences. I have indicated in broad strokes what such perspectival pluralism might look like. The specific details will have to wait for another occasion.

Notes

*The author would like to thank Charles Wolfe for his helpful and insightful comments on an earlier version of this paper.

1. This chapter is based in part on a poster that I presented with Edda Thiels (Department of Neurobiology, University of Pittsburgh) at the 2011 Annual Society for Neuroscience meeting in Washington, DC (Sullivan and Thiels 2011).

2. The members of the group have held several additional meetings, but these meetings have not, as far as a literature search revealed, yielded published proceedings.

3. I think at best what can be said here is that *many* philosophers and social scientists take the intentional stance towards human organisms. However, the extent to which this stance is informed by cognitive scientific thinking about the nature of mind varies across practitioners.

4. The precise aim of the *Decade of the Brain* was, according to G.W. Bush, "to enhance public awareness of the benefits to be derived from brain research" (Presidential Proclamation 3168, 1990). In recent years, as governments and granting agencies have cut funding to scientific research, being able to communicate effectively with members of the general public has become key to generating funds for scientific research.

5. The lack of interest in folk psychology as a viable conceptual-explanatory framework may be regarded as an impediment to the DoM. At a bare minimum, trying to understand the relationship between the *intentional stance*, which common folk and some scientists take towards human organisms, and the *physical stance* – the assumption that an organism's behavior has internal physical (e.g., neural, biochemical) causes – seems prerequisite for effective interdisciplinary communication. No reasons exist to think practitioners in areas of science outside of neuroscience will completely abandon their appeals to folk psychological explanations of behavior, nor is the eliminative materialism for which Paul Churchland (1981) advocates obviously in the offing. Furthermore, given that misunderstandings between neuroscientists and ordinary folk who are looking towards neuroscience for answers may also arise, it seems legitimate for the sake of clarity for neuroscientists to be clear about how they understand the mind and how and in what ways that differs from how non-scientists think about it.

6. As I explain later in the chapter, I think neuroscientists are interested primarily in explaining the capacities of organisms. Recent work in philosophy of neuroscience supports this conclusion (e.g., Bechtel 2008, Bickle 2006, Piccinini and Craver 2011). So, while neuroscientists may be concerned with determining where in the brain *believing*, *wanting*, and *intending* occur, they

are not interested in localizing beliefs, desires, and feelings in the physical world.

7. Clearly, if dispositional analysis is a prerequisite for componential analysis when it comes to cognitive capacities, then proceeding to the explanatory stage that involves componential analysis does require the intentional stance. That this is true is clear in Fodor 1968.

8. To use Fodor's example, "a psychologist might seek to explain failures of memory by reference to the decay of a hypothetical memory 'trace'" (1968, 108).

9. However, as I aim to show in Section 3, the story is a bit more complicated. For psychological explanation will only proceed *successfully* to the second phase in those cases in which an investigator has arrived at the correct functional decomposition, which requires that the intentional stance be adopted in the first phase. Sometimes, however, explanation will proceed to the next stage before one has arrived at the correct functional decomposition, and in the process of investigating the mechanisms of the purported function, investigators will realize they have to "reconstitute the phenomenon" (Bechtel 2008; Craver 2007; Sullivan 2010b). Implicit in my claims in this paper is that reconstituting the phenomenon requires an investigator to consider the intentional stance in addition to considering the physical constitution of the object of inquiry.

10. Piccinini and Craver 2011 argue that a task analysis is nothing more than an incomplete mechanistic explanation of a cognitive capacity. However, to introduce this relationship has the consequence of not keeping investigative strategies, of which task analysis is one, separate from explanatory strategies or models of explanation like mechanistic explanation. Task analysis serves an important function in the context of experimental design – a function it is not obvious that mechanistic explanation can replace.

11. The Stroop task consists of three different types of stimulus conditions that vary across trials. In the *congruent condition*, subjects are presented visually with a word and the color of the text of the word matches the color word (e.g., "red" is presented in red-faced type). In the *incongruent condition* the color of the text differs from the color word (e.g., "red" is presented in green-faced type), and in the *neutral condition* a color-neutral word is presented in either red or green type. The subject's reaction time from the point of presentation of the stimulus on a given trial to the point of responding with the correct word for the color seen (but not read) is measured, and errors in identifying the correct color word are recorded.

12. Watson took this to include introspectionist psychology's "illegitimate children," namely, "functionalist psychology" and "gestalt psychology" (Watson [1924]1970, p. 1).

13. I do not mean here to deny other important critiques of contemporary neuroscience (e.g. Bennett and Hacker 2003), only to point to the irony that the criticisms of behaviorism raised at the Hixon symposium are still applicable to modern cognitive neurobiology (see also Machamer 2009).

14. I want to thank Muhammad Ali Khalidi who, in response to a talk I gave at York University in January 2012, encouraged me to read Wimsatt 2007 as a means to get clear on the kind of pluralism I was advocating.

15. As evidence for this lack of standardization, neurobiologist Yadin Dudai published a dictionary in 2002 in an effort to standardize concepts across those areas of neuroscience that study learning and memory.
16. We see this problem perhaps most clearly in the neuroscientific study of mental disorders in which there has been to date a lack of coordination, particularly between cognitive neurobiological and cognitive neuroscientific approaches (Sullivan forthcoming).

References

Albus, J.S., G.A. Bekey, J.H. Holland, N.G. Kanwisher, J.L. Krichmar, M. Mishkin, S.M. Dharmendra, M.E. Raichle, G.M. Shepard, and G. Tononi. (2007). A Proposal for a Decade of the Mind Initiative. *Science*, 317, 1321.

Barch, D., C. Carter, and J. Cohen. (2004). Factors Influencing Stroop Performance in Schizophrenia. *Neuropsychology*, 18(3), 477–484.

Batra S., J. Sullivan, and D.S. Geldmacher. (2013). Qualitative Assessment of Self-Identity in Advanced Dementia. Poster presented at *Dementia Care at the 13th Alzheimer's Association International Conference: Translating Research to Practice*, Boston, MA.

Bechtel, W. (2008). *Mental Mechanisms: Philosophical Perspectives on Cognitive Neuroscience*. New York: Lawrence Erlbaum.

Bennett, M.R. and P.M.S. Hacker. (2003). *Philosophical Foundations of Neuroscience*. Malden, Mass.: Blackwell.

Bickle, J. (2003). *Philosophy and Neuroscience: a Ruthlessly Reductive Account*. Dordrecht: Kluwer.

Bickle, J. (2006). Reducing Mind to Molecular Pathways: Explicating the Reductionism Implicit in Current Cellular and Molecular Neuroscience. *Synthese*, 151, 411–434.

Bliss T. and Lømo, T. (1973). Long-lasting Potentiation of Synaptic Transmission in the Dentate Area of the Anaesthetized Rabbit Following Stimulation of the Perforant Path. *Journal of Physiology*, 232 (2), 331–56.

Bush, G. (1990). *Presidential Proclamation 6158*, Library of Congress website. (http://www.loc.gov/loc/brain/proclaim.html).

Carter, C., J. Kerns, and J. Cohen. (2009). Cognitive Neuroscience: Bridging Thinking and Feeling to the Brain and Its Implications for Psychiatry. In *Neurobiology of Mental Illness* (pp. 168–178), 3rd edn, ed. D. Charney and E. Nestler. Oxford: Oxford University Press.

Churchland, P.M. (1981). Eliminative Materialism and the Propositional Attitudes. *The Journal of Philosophy*, 78(2), 67–90.

Cohen, J., D. Barch, C. Carter, and D. Servan-Schreiber. (1999). Context-Processing Deficits in Schizophrenia: Converging Evidence from Three Theoretically Motivated Cognitive Tasks. *Journal of Abnormal Psychology*, 108 (1), 120–133.

Craver, C. (2003). The Making of a Memory Mechanism. *Journal of the History of Biology*, 36(1), 153–195.

Craver, C. (2007). *Explaining the Brain: Mechanisms and the Mosaic Unity of Neuroscience*. Oxford: Oxford University Press.

Craver, C. and L. Darden. (2001). Discovering Mechanisms in Neurobiology: The case of Spatial Memory, in P.K. Machamer, R. Grush, and P. McLaughlin (eds),

Theory and Method in the Neurosciences. Pittsburgh: University of Pittsburgh Press.

Cronbach, L. and P. Meehl. (1955). Construct Validity in Psychological Tests. *Psychological Bulletin*, 52, 281–303.

Cronbach, L. and P. Meehl. (1955). Construct Validity in Psychological Tests. *Psychological Bulletin* 52: 281–302.

Cummins, R. (1983). *The Nature of Psychological Explanation*. Cambridge, Mass.: MIT Press.

DeCharms, C.R., and A. Zador. (2000). Neural Representation and the Cortical Code. *Annual Review of Neuroscience*, 23, 613–647.

Dennett, D. (1987). *The Intentional Stance*. Cambridge, Mass.: MIT Press.

D'Hooge, R., and P. De Deyn. (2001). Applications of the Morris Water Maze in the Study of Learning and Memory. *Brain Research Reviews*, 36, 60–90.

Dudai, Y. (2002). *Memory from a to z: Keywords, Concepts and beyond*. Oxford: Oxford University Press.

Fodor, J. (1968). *Psychological Explanation*. New York: Random House.

Fodor, J. (1975). *The Language of Thought*. New York: Cromwell.

Gardner, H. (1985). *The Mind's New Science: A History of the Cognitive Revolution*. New York: Basic Books, Inc.

Gibson, J.J. (1960). The Concept of a Stimulus in Psychology. *The American Psychologist*: 694–703.

Giere, R. (2010). *Scientific Perspectivism*. Chicago: University of Chicago Press.

Hebb, D.O. (1949). *The Organization of Behavior: A Neuropsychological Theory*. New York: Wiley.

Hinde, R.A. (1973). Constraints on Learning: An Introduction to the Problems. In R.A. Hinde and J. Stevenson Hinde, eds *Constraints on Learning* (pp. 1–19). London: Academic Press.

Kandel, E.R., and W.A. Spencer. (1968). Cellular Neurophysiological Approaches in the Study of Learning. *Journal of Physiology*, 48, 65–134.

Kohler, W. (1947). *Gestalt Psychology: An Introduction to New Concepts in Modern Psychology*. New York: The New American Library.

Kuhn, T. (1962). The Structure of Scientific Revolutions. Chicago: University of Chicago Press.

Lømo, T. (2003). The Discovery of Long-Term Potentiation. *Philosophical Transactions of the Royal Society of London B: Biological Sciences*, 358, 707–714.

Lorenz, K. (1965). *Evolution and Modification of Behavior*. Chicago: University of Chicago Press.

Machamer, P. (2009). Neuroscience, Learning and the Return to Behaviorism. In *The Oxford Handbook of Philosophy and Neuroscience*, ed. J. Bickle. Oxford: Oxford University Press.

Perlstein, W., C. Carter, D. Barch, and J. Baird. (1998). The Stroop Task and Attention Deficits in Schizophrenia: A Critical Evaluation of Card and Single-Trial Stroop Methodologies. *Neuropsychology*, 12(3), 414–425.

Piccinini, G. and Craver, C.F. (2011). Integrating Psychology and Neuroscience: Functional Analyses as Mechanism Sketches. *Synthese*, 183, 283–311.

Spitzer, M. (2008). Decade of the Mind. *Philosophy, Ethics, and Humanities in Medicine*, 3, 7–11.

Sullivan, J.A. (2009). The Multiplicity of Experimental Protocols: A Challenge to Reductionist and Non-Reductionist Models of the Unity of Neuroscience. *Synthese*, 167, 511–539.

Sullivan, J.A. (2010a). a Role for Representation in Cognitive Neurobiology. *Philosophy of Science*, 77(5), 875–887.

Sullivan, J.A. (2010b). Reconsidering Spatial Memory and the Morris Water Maze. *Synthese*, 177(2), 261–283.

Sullivan, J.A. (forthcoming). Stabilizing Mental Disorders: Prospects and Problems. In *Classifying psychopathology: Mental kinds and natural kinds*, eds H. Kincaid and J.A. Sullivan. Cambridge, Mass.: MIT Press.

Sullivan, J.A. and E. Thiels. (2011). The Place of the Mind in Contemporary Neuroscience. (Poster presented at the *Society for Neuroscience Annual Meeting* Washington, DC).

Sweatt, J.D. (2009). *Mechanisms of Memory* (2nd edn). London: Academic Press.

Taylor, C. (1964). *The Explanation of Behavior*. London: Routledge and Kegan.

Von Eckardt, B. (1993). *What Is Cognitive Science?* Cambridge, Mass.: MIT Press.

Watson, J.B. (1924 [1970]). *Behaviorism*. New York: W.W. Norton & Company.

Wimsatt, W. (2007). *Reengineering Philosophy for Limited Beings: Piecewise Approximations to Reality*. Cambridge, Mass.: Harvard University Press.

4
Neuroconstructivism: A Developmental Turn in Cognitive Neuroscience?

Denis Forest

4.1 Introduction

Since its birth, brain science has been for the most part the study of the structure and functioning of an already formed brain, the study of the endpoint of a process. Brodmann areas, for instance, are cortical areas of the adult brain (Brodmann 1909). In his authoritative *Neurobiology*, Shepherd devotes only one chapter (out of thirty) to developmental neurobiology (Shepherd 1994). From early attempts at functional localization by Gall or Broca to recent neurocognitive models like the model of visual cognition proposed by Milner and Goodale (Milner & Goodale 2006), functional decomposition of the brain has essentially remained the decomposition of the brain of the adult. Neuroconstructivism, then, as it has been recently vindicated (Mareschal et al. 2007; Sirois et al. 2008) could be understood, first, as the idea that we should take brain development more seriously. This suggestion comes at a time when in many fields of biology, ontogenetic development has become the object of both fascinating discoveries and intense speculation. But there is more to neuroconstructivism than a developmental perspective on the brain, as it can be understood as a view of cognition: it is this view of cognition that motivates a specific, renewed approach to the human brain. What neuroconstructivism is challenging, in fact, is a view of cognitive *explanation*, and of cognitive *development*.

The view of cognitive explanation it rejects is in part derived from the idea that Marr expressed when he said that we should not begin by the study of feathers if we want to understand bird flight (Marr 1982). It is the idea of independent levels of investigation, the idea that psychological explanation is, in principle, fruitfully divorced from the study of low-level implementation. Neuroconstructivism suggests a view of cognitive explanation where there is no point to the separation of levels. There is no such separation in practice because neural events are causally relevant to the understanding of cognitive development in its many forms, both typical and atypical. And there is no such separation in another sense because investigation at different levels (levels being this time: the cell, the brain, the whole body) may rely on the same kind of explanatory factors. As "mechanisms" of the same kind operate at different levels, explanations in different fields are fundamentally of the same type. The unity of cognitive research, then, is not obtained via the existence of a discipline that would shape the whole field (e.g., evolutionary psychology, seen as a guide to neuroscientific research) but through the use of recurring patterns of explanation. And as neuro-constructivism adopts as its highest "core" principle, the principle of "context-dependence,"[1] the brain is not the largest unit, the largest containing system that it considers. Even the developing brain and its changing abilities have to be contextualized. This is why neuroconstruc-tivism perceives itself as the convergence of work in different disciplines and trends of research, namely: developmental neurobiology (the study of "encellment" – see Shepherd 1994, chapter 9); developmental cogni-tive neuroscience (the study of "enbrainment" – see Johnson 2005a); developmental embodied cognition (the study of "embodiment" – see Thelen and Smith 1994).

The view of cognitive development it rejects is the view that cognitive abilities are highly canalized biological features (Ariew 1996), i.e., features that develop in a similar fashion in widely different environments, and are essentially insensitive to environmental variation. In contrast, neuroconstructivism also suggests a view of cognitive development that is highly sensitive to events in both internal and external environments. It rejects both a deterministic view of epigenesis and a view of cognition where inborn abilities would simply unfold in time. Brain development matters, in this sense, not just because there is no pre-existing, detailed blueprint of its organization – because, for instance, the 'protomap' view of the cortex (Rakic 1988), where the identity of any cortical neuron can be traced back to the specific spatio-temporal circumstances of its formation in the proliferative zone, has now been abandoned (Sur and Rubenstein 2005). Brain development matters because we may learn from it, not only about the adult brain, but about how we acquire our own mental powers and what constrains a child's typical or atypical cognitive development.

I do not intend to comment here on every single aspect of the neuro-constructivist program, such as the importance it gives to computational modeling, or its endorsement of the embodied cognition view. My first goal is to provide an understanding of the reasons why neuroscience may need such a program – which gaps it is supposed to fill, which inferences it challenges. I shall therefore focus on early critiques of inference from functional commitment of cortical areas to their inborn specialization (2), and on the link between the controversy over plasticity that arose in the 1990s and the current neuroconstructivist model of brain development (3). My second goal is to examine the core of the program itself, its guiding "principles" – that is, the principle of context-dependence and the idea of level-independent mechanisms (4). I suggest that such a program, as it stands, suffers from both under-determination and over-generalization (5). Because of these defects, and because developmental neuroscience has not been the main concern of philosophers who analyze neuroscience in terms of mechanisms, a dialogue between neuroconstructivism and the philosophy of neuroscience may yield mutual benefits (6).

4.2 The broader context: the Wundt-Munk controversy about the origins of brain function

The anti-nativist stance of neuroconstructivism can be located with regard to its opposition to recent Chomskyan, Fodorian, or massive modularity views about neurocognitive explanation. But it has roots in older debates that are worth mentioning. There are good reasons to prefer a continuist view of the history of cognitive neuroscience over a discontinuist one. Even if the expression "cognitive neuroscience" itself has not been widely used until recently, what this expression refers to may be seen as a research program that originates in the 19th century, and whose development can be understood as the application to the brain of the heuristic strategies Bechtel and Richardson have called decomposition and localization (Bechtel and Richardson 1993). The central idea in this program is that the discovery of brain mechanisms is key to the understanding of the corresponding mental powers. Notable differences in investigation methods, scientific tools, and intellectual background notwithstanding, privileging continuity should not be controversial, as many topics that belong today to cognitive science were treated in the 1880s by scientists like Meynert, Munk, Ferrier, and Jackson. One of the first philosophers to have taken the measure of such a program was Wundt in his *Elements of Psychophysiology* (Wundt

1880). In this monumental work as in related papers, his aim was to make explicit the founding principles of such a science, to sum up its main discoveries, and to offer critical views about some of its central claims. The critical views were meant as a positive contribution to a young domain of investigation. Researchers like Munk, according to Wundt, offer only a mixture of important scientific discoveries and old prejudices. Making these prejudices explicit is a means to an end: clearing the ground for further questions and investigations. In order to achieve this, Wundt sketches what we may call a constructivist answer to nativist views.

Wundt does not reject the principle of localization of function within the brain, a principle that is central to the mechanistic methodology of brain science. But he thinks that the kinds of methods used by neuroscientists may lead them to adopt views that are not, in fact, supported by available evidence. First, the localization of lesions should not be conflated with the localization of functions. Focal lesions in pathological cases are signs of the involvement of brain regions in cognitive tasks, but they do not, by themselves, indicate what the exact nature of this involvement may be. Second, and more importantly, evidence of localized functions is not evidence of an inborn commitment of brain areas to perform definite functions. We should not replace the old holistic error (no functional decomposition of the brain is possible, or fruitful) with what Wundt calls the phrenological error (Wundt 1891): the claim that each part of the brain has its own, immutable, pre-specified function. For Wundt, in order to reject phrenology, it is not enough to give up the entire list of mental faculties as defined by Gall. Quite strikingly, Wundt sees (much before Fodor, but for opposite reasons) that there is more to Gall than a set of old prejudices that everyone has already overcome. There is a mixture of nativism and what we would call modularity (that is, functional specificity of largely autonomous units or powers, units that may be localized in the brain), and this mixture, in 1880, according to Wundt, still provides the framework of much scientific work done in brain science. This point (no sound inference from functional specialization in the adult brain to inborn commitment of the corresponding brain parts) is central to Wundt's proposal. The task of "physiological psychology" (roughly, our cognitive neuroscience), as it is defined by Wundt, is not only to establish what functional localizations are, but more importantly, *where they come from*. Nativist views, according to Wundt, are just the product of our ignorance in this matter. The first part of Wundt's work, then, is to explain why the nativist view of inborn commitment of brain areas should be rejected. To this end, Wundt relies on recent work on neural plasticity. Brain function, when abolished by the destruction of some quantity of brain tissue, may be spontaneously restored during recovery. This is explained, according to Wundt, by other neural elements "taking over" the defective ones. This shows that with different connections, different input/output conditions, functional specificity of brain components may be altered. The possibility of a substitution of one nervous element to another is evidence against strict preexisting definition of specific neural function. This principle of "Ersatz" function itself falls under another, more general principle that Wundt calls the principle of adaptation: "any central element is adapted to its function as it has to perform it more frequently under the pressure of external conditions" (Wundt 1880, Part I, chapter V, 7: General principles of central functions). Plasticity, then, when it is linked to physiological recovery after brain damage, is not a *sui generis* phenomenon: it is just an extreme case of adaptation as defined in this broad sense[2]: in this case, "being progressively adapted" does not mean doing something better with time (improvement of performance, as in a Jamesian definition of plasticity: James 1890) but being able to alter one's pre-existing function under external (contextual) influences. But Wundt does not stop there. He asks: why is this principle of adaptation valid? Where does the power of neural elements to adjust to external circumstances come from? And his answer is: it comes from the lack of specific functional commitment of central, neural elements at birth. If the definition of function is the *result* of the process of brain organization, then it is not *that* remarkable that neural elements may change their functional role in special circumstances during adult life. It is just that no "restriction of fate" during life, no special adaptation leads to a unique, non-reversible functional commitment. Evidence of plasticity brings back the necessity of a scientific explanation of the outcome of epigenetic processes. Nativism, then, is not a solution to the problem of the origin of functional specialization within the brain, it is just a way to ignore it.[3] Plasticity depends on the adaptive power of neural elements, which itself derives from the fact that their standard, domain-specific activity during adult life should be seen as the temporary, reversible product of the changes imposed by brain organization (for instance, input and output conditions) on initial, non-specific response properties of neural elements. The fact that Wundt has little to say about the details of these changes makes his clear, synthetic view of this cluster of important questions even more remarkable.

4.3   The debate on the significance of neural plasticity for the controversy on representational nativism

Just as Wundt once challenged a phrenological view of inborn functional determination, constructivists have challenged a deterministic view of brain development one century later. The intellectual debt of the co-authors of *Neuroconstructivism* (Mareschal et al., 2007) towards Jeffrey Elman, Annette Karmiloff-Smith and the other co-authors of the manifesto *Rethinking innateness* (Elman et al. 1996) is obvious enough. What is needed in order to understand neuroconstructivism is rather to clarify the meaning of the shift in emphasis from the analysis of neural plasticity (considered in 1996 as evidence against representational nativism) to developmental cognitive neuroscience as it flourished in the following years. This can be done through the debate between the philosopher Richard Samuels and the co-authors of *Rethinking innateness* (Samuels 1998). Samuels holds that Elman and his colleagues have misinterpreted the neurobiological facts: neural plasticity does not, contrary to their claims, falsify representational nativism. O'Leary and Sur, in a striking series of experiments, offered evidence that in the brain of mammals, the sensory cortex of a given region may take response properties of another region under exceptional circumstances. Studies have shown that fetal neurons taken from the "visual" area may come to exhibit the organizational and functional properties of the neurons of the somatosensory region where they have been transplanted (O'Leary and Stanfield 1989). Rewiring experiments have also shown that the somatosensory cortex of ferrets, when it is deprived of its normal input, and when it receives visual inputs early in life, may develop response properties that are typical of the "visual cortex" (Sur, Pallas & Roe 1990). Are we to think, then, that parietal regions that are usually described as 'somatosensory areas' are not intrinsically dedicated to the representation of somatic states? According to Samuels, neural plasticity has no such implications. First, innate properties are not necessarily intrinsic properties (that is, properties that are non-relational). If representational properties are extrinsic, rather than intrinsic properties, representational properties of a given set of neurons may both be innately specified and depend on the relation of these neurons to the rest of the cortex and to the environment. In this case, when we are modifying the input conditions of these neurons, it is no surprise that their representational properties are altered. Samuels thinks that constructivists suggest mistakenly that nativists have to adopt an invariance principle such as

The innately specified (representational) properties of a piece of cortical tissue T are invariant under alterations in T's location within the brain and alterations in the afferent inputs to T.

But Samuels thinks that because of the distinction between intrinsic and innate properties, nativists do not have to accept this principle. Moreover, they are, in fact, committed to what he calls Organism nativism rather than Tissue nativism. Tissue nativism is a claim about representational properties of specific brain parts. Organism nativism is a claim about inborn cognitive abilities of whole organisms or people, and it is entirely independent from claims about innate commitment of brain parts or specific localization of function. Even a refutation of tissue nativism (a refutation that, according to Samuels, experiments on plasticity do not provide) would not be a refutation of representational nativism in general.

Let's focus on tissue nativism, as it is linked to the definition of the functions of brain parts, and as Samuels holds that it can be vindicated against Elman's views. First, innate specification of representational properties is difficult to reconcile with some existing neurophysiological data. The crucial role of the activity of visual areas in blind subjects during Braille reading seems to support a view of sensory areas (the "meta-modal organization" of the brain) where they possess a "purpose general" ability to treat incoming sensory signals rather than an inborn commitment to treat one or another kind of such signals.[4] In the case of Braille reading by blind subjects, response properties of the so-called "visual area" are different from what they are in standard cases, without any transplant or re-wiring. In congenitally deaf mice, it has been shown that some neurons of the auditory cortex develop responses to visual and somatosensory stimulation, and that the response properties of other regions, like the visual cortex, are themselves altered (Hunt, Yamoah and Krubitzer 2006). What evidence do we have, then, of an innate specification of the representational properties of the visual or auditory cortex, and why would we prefer this nativist view to the parsimonious alternative of a lack of the inborn functional commitment of sensory areas? Second, we can ask ourselves what it means for the representational property of a given set of neurons to be "innately specified." Extrinsic properties, says Samuels, are constitutive of representational properties. But extrinsic properties are defined *during* epigenesis and as *the result of* the specification of neural paths. In this case, how could representational properties be innately specified *before* the outcome of this epigenetic process or *independently* of it? How could they have a pre-existing, definite *content*? And if they don't, how could they exist at all? What is called the "innate specification of the representational property of visual neurons" seems to be a convoluted way of saying that neurons of the striate cortex have early in life a higher *probability* to receive visual *inputs* than, say, somatosensory inputs. Calling this an "innate representational property" conflates different levels: it re-describes a frequent but non-necessary correlate of the outcome of a *neurobiological* process as what is (at the *representational* or psychological level) 'meant to be.' However, defeating a nativist critique of conclusions drawn from plasticity experiments is one thing; explaining where the typical organization of the brain of mammals comes from (especially the existence of discrete units like cortical areas and their functional specialization) is another. Neuroconstructivism could be understood as the empiricist answer to the objections made by Samuels to the argument from plasticity against innate specification of representational properties of given brain parts. What is needed is not only manipulations that experimentally alter the extrinsic properties of cortical areas, but an understanding of how functional specialization and extrinsic properties are specified during brain development. This is what neuroconstructivism hopes to provide.

4.4 Neuroconstructivism at work

One way of presenting the neuroconstructivist view may be to begin with the contrast between two theories of biological functions in the philosophical literature: one is the etiological view – functions are nothing but effects selected during biological evolution (Neander 1991), the other is the systemic view of Cummins: the function φ of a component x in a system S is its contribution to the explanation of the ability of S to ψ (Cummins 1975). Although it is fairly uncontroversial that in brain science, research is aimed at discovering Cummins functions, and that neuroscience textbooks provide information about the contribution of activities and/or components to larger systems, some still maintain that where there is no history, there is no function (Jacob and Jeannerod 2003), and that we should consider the functions of brain components as products of evolution by natural selection. I don't want to discuss here the merits of these philosophical views, but rather to point out that we have to take into consideration the subtlety of our analysis. For instance, the ability of place cells in the hippocampus to contribute to the formation of maps of the environment (O'Keefe and Nadel 1978) can be seen as (one of) their function(s), and what we mean by that may be that it is a product of a certain evolutionary history where natural selection has played a role. However, the ability of these cells to contribute to the individual's knowledge of *his* environment requires more than the exist-ence of these cells and the corresponding history of the species; it requires a certain kind of individual history where *tokens* of place cells end up coding for specific places. An explanation of *actual* orientation of a given individual will require a causal analysis where specific activations of place cells contribute to the ability of the hippocampus to form a map of *his* familiar environment: no actual orientation is provided by past selection for place cells *as a type*. This means that *different* interesting stories may be told about the origin of functions and that one of them may be about their ontogeny. In this sense, developmental cognitive neuroscience and neuroconstructivism fill a gap. They require that we address additional questions that are left unanswered by the evolutionary perspective, and that Cummins' style of functional analysis is not meant to solve either:

Question 1: in a given system S, in virtue of what does a component x receive its own power to φ?

Question 2: how did S become able to produce its own characteristic output? Why ψ (rather than some other activity) in S?

Developmental Cognitive Neuroscience has to provide answers to these related questions, to give a mechanistic explanation of how components receive the distinctive powers they have. And the neuroconstructivist idea is that, if we cannot be satisfied with a view where brain develop-ment is just brain maturation, we should adopt what Johnson calls the *Interactive specialization view*. To know why component C has received its distinctive role (question 1), we have to look at the developmental history of the System in which C is embedded and at the corresponding external environment. This developmental history causally explains, in particular, the current pattern of connectivity of C and its response properties. To the second question, how does S become able to ψ?, the neuroconstructivist answer is: we can explain the emergence of S's char-acteristic output ψ through its own interactive specialization and the interactive specialization of its parts. Now to explain how specialization occurs, constructivism offers a set of *domain-general and level-independent* "mechanisms":

Competition: for any given function, neural systems evolve from wide-spread, a-specific activity into specialized correlates

Cooperation: which is another word for functional integration

Chronotopy: key aspects of development rely on sequences of events that are closely related: in particular early specialization of a component

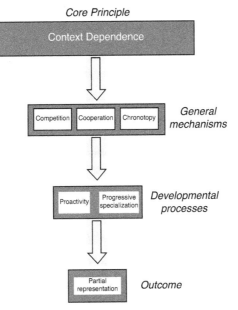

Sirois et al.: Précis of *Neuroconstructivism*

Figure 4.1 The core principle of the Neuroconstructivist Program
Source: Sirois et al. 2008.

A constrains the posterior developmental trajectory of a related component B; explaining the latter is impossible without explicit reference to the former.

Let's take, for instance, face recognition as the *explanandum*. Instead of considering the fusiform area as a cognitive module with an inborn, domain-specific commitment to process face representations, the neuro-constructivist framework invites us, on the cognitive level, to pay attention to the difference between early sensitivity to face-like visual patterns and later development of face recognition itself, a development that may take advantage of an early emerging ability for a-specific (domain-general) visual expertise. It invites us to consider the difference between a sub-cortical route responsible for face detection and a cortical network, of which the fusiform gyrus is a part, involved in face identification (Johnson 2005b): in this case, the specialization for faces of the fusiform area is the product of its interactions with the sub-cortical route and the constant exposure to faces in the social environment. Competition, cooperation of brain components and chronotopy are jointly responsible for the emergence of face recognition during development as a deeply entrenched cognitive feature. According to its proponents, this view is able to account for several phenomena: a) widespread brain activation in young subjects, in response to faces (when specialization through competition has not yet occurred), that contrasts with specific local activation in elder subjects (Scherf et al. 2007); b) the recruitment of the fusiform gyrus in tasks of visual recognition of non-face stimuli by experts in a given domain (Gauthier, Skudlarski, Gore, and Anderson 2000); and c) atypical developmental trajectories, where defects of the sub-cortical processing of faces have cascading effects on other cognitive abilities and later phases of development (Johnson 2005b). Neuroconstructivist explanations of this kind may be considered as a sub-type of what has been called by philosophers of science like Carl Hempel and Ernst Nagel *genetic explanations*. According to Hempel, a genetic explanation "presents the phenomenon under study as the final stage of a developmental sequence, and accordingly accounts for the phenomenon by describing the successive stages of that sequence" (Hempel 1965, 447). This is still an appropriate description of the neuroconstructivist proposal, even if neuroconstructivist explanations do not fit Hempel's covering-law model of scientific explanation. "Mechanisms" like cooperation, competition and chronotopy are proposed to identify causally relevant factors (Craver 2007) through their abstract, generic description, not merely regular sequences of events.

A view of brain development that is context-dependent, in this sense, may not only be accurate, but prove crucial for neuroethical issues. Some studies have shown a marked disadvantage for children of low economic status in tasks involving the prefrontal executive system, the left peri-Sylvian language system, and the medial temporal memory system (Farah, Noble and Hurt 2006). Potential causes range from prenatal substance exposure to nutritional factors (resulting in iron deficiency anemia), effects of environmental stress (the release of hormones that have a negative impact on hippocampal development, for instance) and lack of cognitive stimulation. Maturational views of neurocognitive development that suggest only a triggering role for environmental factors may seriously underestimate the impact of these factors not only on neurobiological, but also on cognitive development. Accordingly, neuroconstructivism may be important at two levels. First, from a theoretical point of view, it sides with an interactive view of individuation, where abilities supervene on the interactions of individuals and their environment during development. Neuroconstructivism may stimulate research on the nature and extent of such interactions. On a practical level, developmental neurobiology may give precious information on how, in matters of public health, we may become able to implement our norms of justice and fairness when it comes to child development.

4.5 Universal context-dependence? A critique

There is a close link, in the neuroconstructivist framework, between an *empirical* claim about neurobiological development and its central *theoretical* claim. The empirical claim concerns the relevant factors of an explanation of cortical development; it says that we should downplay the importance of genetic factors in such an explanation. In favor of that claim, neuroconstructivists offer two main reasons. The first is the role during development of epigenetic factors, in particular, activity-dependent mechanisms and adaptations like the ones that are crucial to the definition of ocular dominance columns (Mareschal et al. 2007, 21). The second is that there is no region-specific gene expression in the cortex: one notable exception is the H-2Z1 transgene that is expressed in only one region in mice, the layer IV of its somatosensory cortex (O'Leary and Nakagawa 2002, 22). Accordingly, neuroconstructivists hold that, although genes are involved in an early definition of broad regional differences, later specification of well-defined areas is mostly the product of activity-dependent processes (Mareschal et al. 2007, 22). This empirical claim is offered as evidence by neuroconstructivists for the validity and heuristic value of the core principles of their model: universal context-dependence, interactive specialization of brain components governed by competition, and cooperation.

"Context-dependence" is often understood in terms of dependence on interactions with the environment, internal but also external. One risk here is overgeneralization: from an evolutionary perspective, while it is obvious that both the level and the type of related activity may be crucial when it comes to the size or the function of a given area (Hunt, Yamoah and Krubitzer 2006), it is hardly obvious that the global cortical architecture can be understood in terms of context-dependence. Comparison between species shows that the global organization of the mammalian cortex (its "Bauplan") is fairly conservative, differences in behavior notwithstanding (Krubitzer 2007). For instance, sensory cortical fields are not context-dependent to the point that blind species like mole rats would be entirely deprived of visual structures: the architectural pattern remains strictly constrained and, to a large extent, context-independent.

To take another example, recent work devoted to ocular dominance columns in primary visual cortex (Crowley and Katz 2002) suggests that their emergence predates the first months of life, and, as a consequence, cannot be the result of activity-dependent competition between thalamic inputs resulting from retinal stimulation. Although the formation of such columns may not be entirely independent from activity, as recent work on the role of retinal waves has shown (Torborg and Feller 2005), what is crucial in this case is endogenous activity triggered by internal factors rather than actual visual experience.

Concerning epigenetic factors in general and activity-dependent change, some similar lessons could be drawn from the phenomenon of axon guidance. Constructivism since the days of Elman and his co-workers (Elman 1996, 245) has argued that we have to make a distinction between additive and subtractive events, initial proliferation of synaptic connections during development and a later phase of "pruning" that corresponds to the degeneration of non-functional paths. This conforms with what is predicted by the epigenetic, "specialization through competition" model. But neural pathways do not develop in a purely anarchic manner before a negative phase of selective apoptosis and degeneration driven by competition. In particular, projection from thalamic regions to the isocortex may largely depend on patterns of regional expression of molecules that function as guidance cues (both positive and negative) for neurite outgrowth. For instance, expression of ligand Ephrin-2A5 in the somatosensory cortex inhibits projection from limbic thalamic afferents (Gao et al. 1998). As expression of these molecules happens early in development, and *precedes* the invasion of the cortex by thalamic axons, it may be viewed as a context-*independent* factor of regional differentiation. This involves something that is very different from the predictions of the proliferation-and-pruning-model. The risk, then, is to neglect explanatory factors that do not fit the model, such as the ones suggested by the pioneering work of Sperry (Sperry 1943) and his idea of chemo-affinity as a factor of organization during development. To sum up, from genuine instances of activity-dependence and thalamic influence we cannot conclude to their explanatory relevance in any given context. And although neuro-constructivists are right to distinguish between activity-dependence and (external) context-dependence (Mareschal et al. 2007, 32) not only does spontaneous endogenous activity have a role where sensory experience has none for the establishment of visual circuitry, but it seems that we also have to take into account activity independent factors (Huberman, Feller and Chapman 2008).

Concerning the second empirical claim, the one concerned directly with genes, even if it remains true that there is no one-to-one correspondence between genes and cortical areas, and even if knowledge in these matters is still fragmentary and based mainly on studies that focus on a single species (mice), not only has evidence of genetic control of arealization been growing in the last twelve years, but this may be considered as the main recent event in the field of the neurobiology of cortical development (O'Leary and Sahara 2008). For instance, gene Emx2 is normally expressed in low rostral to high caudal and low lateral to high medial gradients, and cadherin Cad 8 is a special attribute of motor cortex situated in the rostrally located motor cortex. In Emx2 homozygous mutant mice, however, it has been discovered that the pattern of cadherin expression is markedly altered, expression of Cad8 being expanded both caudally and medially while caudal areas contract. Moreover, in Emx2 mutant mice, connections between cortical areas and thalamic nuclei are significantly altered, arguably because of the involvement of Emx2 in the differential production of molecules controlling axon guidance: while in wild-type mice, the anterior occipital cortex receives projections from the dorsal lateral geniculate nucleus, which conveys visual inputs, in mutant mice, the same region receives projections from the ventroposterior nucleus which are normally characteristic of the somatosensory cortex, a clear sign of the contraction of the visual area (Bishop, Goudreau and O'Leary 2000). Accordingly, as molecular expression and patterns of connectivity are two of the most important attributes of cortical areas, we may conclude that genetic control, in some species at least, goes much further than the rough preliminary definition of whole regions whose internal architecture would be fine-tuned under the influence of epigenetic factors. Moreover, the choice is not between a one-to-one correspondence between genes and cortical areas and no genetic mediation of cortical development whatsoever: genes like Emx2, Pax6, and COUP-TFI are expressed in the cortex according to gradients that may overlap in such a manner that taken together, they play a crucial role in the definition of the emergence of the combination of features that is unique for each area (Kingsbury and Finlay 2001).However suspicious we may be, then, of the metaphor of traits being directly coded or represented in sequences of DNA basis, we do not have to conclude from the fallacy of outdated genetic determinism to the validity of rival constructivist proposals when it comes to the explanation of cortical development. Explanations of arealization now begin with patterning centers contained in the dorsal telencephalon of the developing brain (O' Leary and Sahara 2008). These patterning centers secrete a series of molecules (like the fibroblast growth factor FGF8) which are, in turn, responsible for the differential expression of genes like Emx2, Pax6, and COUP-TFI in progenitor cells and their progeny in cortical regions. This proposal deserves several comments. First, it could be said with reason that expression of genes in this model is context-dependent, but this kind of dependence has to be understood in the perspective of a "regulatory hierarchy" (O'Leary and Sahara 2008) that secures the emergence of a quite uniform and highly adaptive cortical structure. It is not "horizontal" interaction between equals (genes, cells), but hierarchical control that seems to matter the most. Second, even if constructivists are right to insist on the importance of the timing of events during development ("chronotopy"), it seems difficult to understand what this timing depends on without reference to the above-mentioned regulatory hierarchy and, for instance, to early secretion of signaling molecules on which gene expression is dependent. Third, in agreement with an influential view of mechanistic explanation (Machamer, Darden and Craver 2000), explanation of cortical development is not achieved by pointing exclusively to low-level, "bottoming-out" components of cortical structures, but through the careful description of the integration of entities and activities located at different levels of the mechanism. We have to adapt this model in a developmental context: "mechanism" does not refer here to a static set of components, but to a self-modifying structure where interactions are responsible for the addition of new features and operations. But the lesson remains: explanation of phenomena at higher levels of mechanism is neither reducible to bottoming-out entities and activities, nor divorced from them. To sum up, even if the shift from gradients of gene expression to the abrupt contrast between discrete cortical areas is not fully understood, it seems that there is no mechanistic explanation of cortical development without a reference to this regulatory hierarchy of which gene expression is a key part, a proposition that is not easy to reconcile with the spirit of the neuroconstructivist program which favors "horizontal" interactions and epigenetic factors.

The neuroconstructivist program aims to make explicit "lessons" from past studies that "should help us to identify the relevant questions, factors and variables that will lead us to a deeper understanding of development" (Mareschal et al. 2007, 91). It does not seem that its "principles" are to be understood as statements of universal laws of nature; it is nowhere said, for instance, that *comparative* developmental neurobiology would support claims of necessity or universality in these matters. What is offered by neuroconstructivists seems rather to be heuristic principles that may function as guidelines for future research.

However, neuroconstructivist principles may suffer, first, from overgeneralization: it may be that what is valid and heuristically useful for late stages of development and mid-level organization may not have the same value for earlier phases and/or lower levels of organization (see Kingsbury and Finlay 2001, and their distinction between "early cortical regionalization" and "late cortical regionalization"). Second, these same principles may suffer from under-specification, as was already noticed, not only because specialization may be understood in more ways than one (Anderson 2008) but because of the wide differences between types of context-sensitivity. Unless we define unambiguously a) which degrees of change are significant enough to be counted as evidence of sensitivity (that is, as a mark of dependence), b) what is evidence of causal dependence to context, and above all c) what is exactly the context (with its specific boundaries and properties) a given event is supposed to depend on in a given case, it will be very difficult to establish what exactly confirms or disconfirms the principle of context-dependence. Moreover, the kind of interactions between components we have to understand is not always the one that is predicted by neuroconstructivism: often, context-dependence involves hierarchical control within a multi-level developmental mechanism, rather than cross-talk among equals at a given level.

4.6   Conclusion

Very often, constructivism is perceived and debated as an alternative to nativism, as if arguments and empirical predictions could lead to a final settlement of the dispute. It is reasonable to think, however, that scientific investigation itself is inherently pluralistic, that developmental neuroscience as understood by neuroconstructivists, evolutionary neuroscience, and "systemic" neuroscience asks, to use van Fraassen's terminology, different "why-questions," the topic of each of them being associated with a different "contrast class" (van Fraassen 1980). One of the main interests of the developmental perspective is that atypical development is not necessarily synonymous with dysfunction, impairment, and cognitive failure, as was shown by recent work on high-level autism (Happé 1999). But the classical framework of neuropsychology, with its contrast between brain or cognitive integrity and deficits associated with lesions, if not relevant to the field of developmental syndromes, is still valid in its proper context. Neuroconstructivism may (and does) inspire quite promising research (Rippon, Brock, Brown, and Boucher 2007) with valuable theoretical and social implications, but different perspectives are still needed. Pluralism, as it is advocated here, does not preclude crosstalk between existing disciplines and the birth of new integrating disciplines at their borders (evolutionary developmental neuroscience would be an example). But it precludes seeing a proposal like neuroconstructivism and the new emphasis on development in terms of "developmental turn" or paradigm shift.

Philosophers interested in neuroconstructivism may face the following alternative. One possibility is to use the neuroconstructivist framework to build a broad view of development and human nature. The other possibility is to reflect on the program itself and its current limitations. The second possibility defines one possible task for the philosophy of neuroscience. On the one hand, if mechanisms have been the focus of attention in recent years in the field of philosophy of neuroscience, the developmental perspective is an occasion to consider these mechanisms in a different light: mechanisms and their characteristic activities are not only the producers of change, they are also the products of change, something we need to understand if we want to know how they become capable of doing what they do. On the other hand, the core idea that mechanisms typically span multiple levels (Machamer, Darden and Craver 2000; Craver 2007) is still fruitful in this different context, as it cures us both from strict fundamentalism (only lower levels matter) and vague emergentism. In particular, it is only in defining the role of genes in the containing systems where they are embedded, and it is only in considering the phenomenon of arealization in its relation with its tight but complex genetic control, that developmental neuroscience will be able to overcome the present limitations of the rhetoric of 'construction.'

Notes

1. Mareschal et al. 2007, chapter 5; Sirois et al. 2008, p. 325.
2. A strikingly similar view is expressed in Elman et al., 1996, pp. 247–248.
3. For instance, Wundt sees very clearly the implications of a constructivist view for a question like the "specific nervous energy" inherited from Müller, a problem that had become in his day that of the relation between sensory *qualia* and brain regional activation. For him, there is no individualistic, internalist explanation of the differences between sensory modalities. The supervenience base of visual experience cannot be the activation of the "visual" region of the brain alone, independently from the nature of sensory input.
4. Pascual-Leone and Hamilton, 2001. The meta-modal organization hypothesis takes a middle ground in the debate on domain-specific or domain-general abilities of neuro-cognitive systems. It does not involve any claim of

equipotentiality of brain regions, but rejects an inference from specialization of a sensory region to a specific domain to an inborn commitment to this specific domain.

References

Anderson, M.A. (2008). Are Interactive Specialization and Massive Redeployment Compatible?, *Behavioral and Brain Sciences*, 31, 331–334.

Ariew A. (1996). Innateness and Canalization. *Philosophy of Science*, 63, 19–27.

Bechtel, W., and Richardson, R. (1993). *Discovering Complexity, Decomposition and Localization as Strategies in Scientific Research*. Princeton: Princeton University Press.

Bishop, K.M., Goudreau, G., and O'Leary, D. (2000). *Emx2* and *Pax6* Regulate Area Identity in the Mammalian Neocortex. *Science*, 288, 344–349.

Brodmann, K. (1909/1994). *Localisation in the Cerebral Cortex* [translation of *Vergleichende Lokalisationslehre der Grosshirnrinde*]. London: Smith-Gordon.

Craver, C. (2007). *Explaining the Brain*. Oxford: Oxford University Press.

Crowley, J.C. and Katz, L.C. (2002). Ocular Dominance Revisited. *Current Opinion in Neurobiology*, 12(1), 104–109.

Cummins, R. (1975). Functional Analysis. *Journal of Philosophy*, 72, 741–764.

Elman, J., Bates, E.A., Johnson, M.K., Karmiloff-Smith, A., Parisi, D., and Plunkett, K. (1996). *Rethinking Innateness*. Cambridge, Mass.: MIT Press.

Farah, M., Noble, K.G., and Hurt, H. (2006). Poverty, Privilege, and Brain Development. In J. Illes ed., *Neuroethics*, pp. 277–287. Oxford, Oxford University Press.

Gao, P.-P., Yue, Y., Zhang, J.-H., Cerretti, D., Levitt, P., and Zhou, R. (1998). Regulation of thalamic neurite outgrowth by the Eph ligand ephrin-A5. *Proceedings of the National Academy of Science*, 95, 5329–5334.

Gauthier, I., Skudlarski, P., Gore, J.C., and Anderson, A.W. (2000). Expertise for Cars and Birds Recruits Brain Areas Involved in Face Recognition. *Nature Neuroscience*, 3(2), 191–197.

Happé, F. (1999). Autism: Cognitive Deficit or Cognitive Style? *Trends in Cognitive Sciences*, 3(6), 216–222.

Hempel, C. (1965). *Aspects of Scientific Explanation*, New York: Free Press.

Huberman, A., Feller, M., and Chapman, B. (2008). Mechanisms Underlying Development of Visual Maps and Receptive Fields. *Annual Review of Neuroscience*, 31, 479–509.

Hunt, D.L., Yamoah, E.N., and Krubitzer, L. (2006). Multisensory Plasticity in Congenitally Deaf Mice: How Are Cortical Areas Functionally Specified? *Neuroscience*, 139, 1507–1524.

Jacob, P. and Jeannerod, M. (2003). *Ways of Seeing: The Scope and Limits of Visual Cognition*. Oxford: Oxford University Press.

James, W. (1890). *The Principles of Psychology*. New York: Holt.

Johnson, M. (2005a). *Developmental Cognitive Neuroscience*, 2nd edn. Oxford: Blackwell.

Johnson, M. (2005b). Subcortical Face Processing. *Nature Reviews Neuroscience*, 6(10), 766–774.

Kingsbury, M.A., Finlay, B.L. (2001). The Cortex in Multidimensional Space: Where Do Cortical Areas Come From? *Developmental Science*, 4(2), 125–142.

Krubitzer, L. (2007). The Magnificent Compromise: Cortical Field Evolution in Mammals, *Neuron*, 56, 201–208.

Machamer, P., Darden, L., Craver, C. (2000). Thinking about Mechanisms. *Philosophy of science*, 67(1), 1–25.

Mareschal, D., Johnson, M.J., Sirois, S., Spratling, M.W., Thomas, M.S., and Westermann, G. (2007). *Neuroconstructivism: How the Brain Constructs Cognition.* Oxford: Oxford University Press.

Milner, A.D. and Goodale, M.A. (2006). *The Visual Brain in Action.* Oxford: Oxford University Press.

Marr, D. (1982). *Vision.* New York: Freeman and Co.

Neander, K. (1991). The Teleological Notion of "Function." *Australasian Journal of Philosophy*, 69(4), 454–468.

O'Keefe, J. and Nadel, L. (1978). *The Hippocampus as a Cognitive Map.* Oxford: Oxford University Press.

O'Leary, D.D. and Nakagawa, Y. (2002). Patterning Centers, Regulatory Genes and Extrinsic Mechanisms Controlling Arealization of the Neocortex. *Current Opinion in Neurobiology*, 12(1), 14–25.

O'Leary, D.D. and Sahara, S. (2008). Genetic Regulation of Arealization of the Neocortex, *Current Opinion in Neurobiology*, 18/1, 90–100.

O'Leary, D.D. and Stanfield, B.B. (1989). Selective Elimination of Axons Extended by Developing Cortical Neurons Is Dependent on Regional Locale: Experiments Utilizing Fetal Cortical Transplants. *Journal of Neuroscience*, 9, 2230–2246.

Pascual-Leone, A. and Hamilton, R. (2001). The Metamodal Organization of the Brain. In Casanova, rC. and Ptito, M., *Progress in Brain Research*, 134, 1–19.

Rakic, P. (1988). Specification of Cerebral Cortical Areas. *Science*, 241, 170–176.

Rippon, G., Brock, J., Brown, C., and Boucher, J. (2007). Disordered Connectivity in the Autistic Brain: Challenges for the New Psychophysiology, *International Journal of Psychophysiology*, 63(2), 164–172.

Samuels, R. (1998). What Brains Won't Tell Us about the Mind: a Critique of the Neurobiological Argument against Representational Nativism. *Mind and Language*, 13(4), 588–597.

Scherf, K.S., Behrmann, M., Humphreys, K., and Luna, B. (2007). Visual Category-Selectivity for Faces, Places and Objects Emerges along Different Developmental Trajectories. *Developmental Science*, 10(4), 15–30.

Shepherd, G.M. (1994). *Neurobiology.* Oxford: Oxford University Press.

Sirois, S., Spratling, M., Thomas, M.S., Westermann, G., Mareschal, D., and Johnson, M.H. (2008). Précis of Neuroconstructivism: How the Brain Constructs Cognition. *Behavioral and Brain Sciences*, 31, 321–331.

Sperry, R. (1943). Effect of a 180 Degree Rotation of the Retinal Field on Visuomotor Coordination. *Journal of Experimental Zoology*, 92, 263–279.

Sur, M. and Rubenstein, J.R.L. (2005). Patterning and Plasticity of the Cerebral Cortex. *Science*, 310, 805–810.

Sur, M., Pallas, S.L., and Roe, A.W. (1990). Cross-Modal Plasticity in Cortical Development: Differentiation and Specification of Sensory Neocortex. *Trends in Neuroscience*, 13, 227–233.

Thelen, E. and Smith, L.B. (1994). *a Dynamic Approach to the Development of Cognition and Action*. Cambridge, Mass., MIT Press.

Torborg, C.L. and Feller, M.B. (2005). Spontaneous Patterned Retinal Activity and the Refinement of Retinal Projections. *Progress in neurobiology*, 76(4), 213–235.

van Fraassen, B.C. (1980). *The Scientific Image*. Oxford: Clarendon Press.

Wundt, W. (1880). *Grundzüge der physiologischen Psychologie*, 2nd edn, Leipzig, Engelmann.

Wundt, W. (1891). Zur Frage der Localisation der Grosshirnfunctionen. *Philosophische Studien*. Leipzig: Wilhelm Engelmann Verlag.

Part II
Naturalistic Approaches

5
Computing with Bodies: Morphology, Function, and Computational Theory

John Symons and Paco Calvo

Contemporary philosophers inherit an anti-psychologistic tradition. The central figures in the early history of both the continental and analytic movements opposed what they saw as the encroachment of psychologists and their fellow travelers on the territory of philosophers (see Kusch 1995 and Dummett 1993).[†] Most prominently, both Frege and Husserl argued that we should avoid corrupting the study of thought with psychologism. As they understood it, psychologism is the view that the best way to understand thought is to look to the empirical study of what we (or our brains) happen to do when we're thinking. Thought itself, on their view should be understood apart from the empirical investigation of mind, let alone the study of the gory details of the brain and nervous system.

The emergence of the computational model of mind in the 20th century, and specifically the conceptual distinction between structural and functional properties of computational systems seemed to provide a non-reductive account of mind, and computationalism proved quite compatible with the anti-psychologistic tendencies of mainstream analytic philosophers. Treating pain, belief, desire, etc. as functionally individuated concepts allowed philosophers to resolve the tension between anti-psychologism and a commitment to the progress of empirical science. Computationalism maintained the autonomy of mental properties as functional states of a physical system, thereby protecting our commonsense understanding of mental life from revision at the hands of the empirical sciences. Since computational functionalism is usually presented as a non-reductive *physicalist* theory, adherents get autonomy for their account of mental life, without being forced to commit to what they regarded as a metaphysically problematic form of ontological dualism. In this way, computationalism supports autonomy without abandoning broader scientific principles. This, at least, is the way the story is usually told.

In this paper we examine some of the implications of our understanding of the body for computational theory. We shall argue that one of the most important constraints on computational theorizing in the study of mind is the initial determination of the challenges that an embodied agent faces. One of the lessons of an area of robotics known as morphological computation is that these challenges are inextricably linked to an understanding of the agent's body and environment. The simple conceptual point is that computational theory is part of the science of mind insofar as it serves to answer questions about the kinds of minds that we and other creatures have. David Marr, one of the most important early figures in Cognitive Science and a pioneering figure in the computational theory of vision, provided an apt summary of the questions that computational theory works to answer:

> Computational Theory: What is the goal of the computation, why is it appropriate, and what is the logic of the strategy by which it can be carried out? (Marr 1982, 25)

Contrary to the way Marr's view of computational theory is sometimes read, and contrary to many of his other methodological claims, we suggest that these questions point to the study of the body and the environment as indispensable parts of computational theorizing in the science of mind.[1] Answering these questions, we argue, requires a detailed understanding of the agent's body and environment. We differ with traditional computationalist philosophers of mind with respect to the starting point of inquiry. On our view, computational theorizing must begin with an understanding of the challenges faced by the agent. It also involves understanding how the morphological characteristics of the body can serve as solutions to computational problems. Knowing how the body solves problems will be part of what a computational theory tells us about an agent.

Taking this approach to computational theory allows a more precise treatment of the kinds of insights that have been presented by advocates of embodied theories of cognition. For the most part, when such theories figure in philosophical debates, they have had relatively little scientific content. This paper offers a way of understanding how philosophical arguments for the importance of the body and environment can be translated into scientific questions for a computational theory of mind. As we argue here, no good computational theory of mind ignores the body and environment. Pace Hilary Putnam, if we were to discover that we are made of Swiss cheese, it actually would matter for the science of mind.

5.1 Anomalous monism, ordinary language philosophy, and conceivability arguments

In this section we briefly review three lines of argument that run counter to our central claim before turning to morphological computation in detail. We regard each of these arguments as flawed and have addressed some of them directly elsewhere. We do not claim to decisively refute these arguments here, but present them simply in order to explain to non-philosophical audiences why it is even necessary to defend the importance of embodiment in computational theory.[2]

In contemporary philosophy of mind, resistance to the empirical study of the brain and body comes in varying degrees and is motivated by three basic kinds of argument. All of these assume that some protected domain of propositions concerning the mind is not subject to revision. The scope and limits of this insulated domain differs from thinker to thinker, and there may always be ways to construct gerrymandered or increasingly anemic conceptions of mental life that can be protected from revision. Our goal is not to block the possibility that there might be something that is not subject to revision via empirical investigation. Instead, we will begin by examining some of most prominent anti-empirical strategies before comparing them with a revised model of computational theory that takes biological, and more specifically morphological, considerations seriously.

Donald Davidson's well-known anomalous monism arguments (see Davidson 1970 in particular) have been understood to license the claim that psychology has no intrinsic relationship to the so-called 'lower-level' sciences. These arguments have led some philosophers to criticize, in principle, any attempt to connect neuroscientific inquiry with the investigation of mental life. The trouble with anomalous monism is that it seems to leave psychological phenomena disconnected from the causal economy of the physical world. Davidson attempted to answer the charge that contrary to his stated views, anomalous monism leaves mental life epiphenomenal (See for example Davidson 1993), but his later defense of anomalous monism has not satisfied critics. (See for example McIntyre 1999.) Nevertheless, in spite of their apparently unpalatable consequences, Davidson's arguments for the autonomy of psychology have been highly influential among philosophers.

A less influential line of argument derives from ordinary language philosophy. For example, in their *Philosophical Foundations of Neuroscience* (2003), Bennett and Hacker argued that neuroscience is motivated by faulty reductionist presuppositions and simple logical fallacies. Following Wittgenstein, they argue that scientists are simply mistaken when they claim that brains believe, interpret, decide, etc. On their view, only human beings are capable of doing such things. Brains are not. Their sweeping critique of contemporary neuroscience rests on the idea that ascribing information processing capacity to brains involves a confusion of a part (the brain) for a whole (the person). They call this the mereological fallacy.[3] Bennett and Hacker assert that the brain cannot be treated as a possible subject of belief and desire and there is nothing that the brain does the can be predicted on the basis of its beliefs and desires. This contention represents the most extreme form of those views that seek to insulate our understanding of the mind from possible revision. The heart of this argument seems to be something like the following: Since we don't ordinarily talk about brains as having beliefs, desires, and the like, any such talk has violated the norms governing the use of these terms. Since the source of the norms governing the use of the terms is ordinary usage, deviant usage, or attempts to revise that usage, are simply misuses or misunderstandings of those terms. Such a position is implausibly conservative insofar as it seems to assume that ordinary usage fixes the meanings of terms in ways that do not permit revision in light of future evidence. The history of science is replete with examples that would undermine this faith in the authority of ordinary usage.

Conceivability arguments have served as a third strategy for attempting to set the mind apart from the rest of the natural world by reference to one or another of its allegedly essential characteristics. According to proponents, these essential characteristics are *conceivably* separable and therefore *possibly* separable. If they are possibly separable, then they are not necessarily identical to the brain and body. Since ontological identity must be necessary identity, these properties of the mind cannot be identical to the brain. So, for example, many philosophers are convinced of the ontological peculiarity of phenomenal experience because of what they take to be the impossibility of necessary a posteriori identity statements linking minds and bodies. Saul Kripke provided the crucial components of this argument in *Naming and Necessity* (Kripke 1980, 148–155).

While these three lines of argument differ, they all support the view that biological or other empirical evidence can have no direct bearing on those aspects of the mind that they regard as autonomous. Of course, as John Bickle points out, even the most hardcore "autonomist" will concede that some aspects of mental life admit of a biological explanation (Bickle 1998, 1).[4] When philosophers assert the autonomy of the mental, they are usually excepting only a subset of the phenomena we ordinarily associate with mental life from reduction or revision. What anti-neuroscientific philosophers take this genuinely cognitive or phenomenal subset to be varies from thinker to thinker. For Putnam, for example, reason's normative powers fall beyond the purview of neuroscience, while for Fodor and others, our ordinary folk psychological talk of belief, desire, and action stands as an inviolable and unrevisable benchmark for all claims about the mind. For Chalmers, the content of phenomenal judgments will never be reducible to biology, etc. (1996), for these philosophers, biological mechanisms can safely be ignored insofar as their particular sub-domain of the mental is, for one reason or another, autonomous.

While this paper cannot respond directly to all the anti-biological arguments in the literature, our goal is to critically examine the idea that the concepts that we use to talk about mental life – at least when those concepts are characterized in a computational theory – can legitimately float free of our knowledge of the body and the environment. Insofar as computational theory treats the mind in functional terms, it seems like functions should float free of the body. For the most part, the functions that we associate with mental life can (at least in principle) be implemented by a variety of structures – this is the nature of functions. Putnam, the most important philosopher in the early days of computational functionalism, famously emphasized this point when he claimed that as far as our study of mental life is concerned "we could be made out of Swiss cheese and it wouldn't matter" (Putnam 1975, 291). From the perspective of a computational functionalist like Putnam, too much emphasis on neuroscience and embodiment run the risk of confusing the mind's 'software' with its 'hardware.' As we noted above, the hardware-software and structure-function distinctions have held great appeal for philosophers. In his classic papers from the early 1960s, Putnam was one of the first to identify mental life with a set of functions that are contingently implemented in human brains. Mental states, he claimed, are functional states that just happen to be implemented by biological systems but that could just as easily be implemented in a suitably organized machine, cloud of gas, or anthill.

On our view, the most charitable interpretation of this kind of functionalism is that it is premature. If the Swiss cheese approach made sense, then we would have a finished computational theory of mind. However, given that we don't have a finished theory of mind, then we cannot accept the Swiss cheese view, and we will need to attend to the constraints of the body and the environment in order to take the first step towards a functional account.[5] On our view computational theory needs to begin with an understanding of the problems that the mind must solve before we can claim to have a functional characterization of the mind.

Of course, we are free to stipulate that the concepts we use to talk about mental life are descriptively adequate; that they correctly capture the relevant features of their subject matter. This latter assumption would certainly help, a priori, to make them invulnerable to revision. While many philosophers have held the view that our commonsense folk psychology is perfect just the way it is, we take this to be an implausibly strong assumption for healthy and ongoing inquiry in the science of mind. Rather than assuming that we have the story of mind perfectly well under control, we suggest that, at least for the time being, one ought to be committed to the possibility of inquiry into the nature of mind. If we are correct, then given the questions that computational theory should answer, the idea that our psychological concepts are unrevisable is highly doubtful.

5.2   Morphological computation

So, how should computational theory take account of the properties of the body? Thankfully, we do not have to rely on thought experiments to guide our reasoning here. For the remainder of this paper, we discuss an emerging subfield of robotics known as morphological computation. Work in this area has implications for cognitive science and philosophy of mind. In particular, debates concerning the extended mind and embodied cognition can be brought into sharper focus by attending to some relatively simple cases in the morphological computation literature.

One obvious methodological consequence of this work is that it challenges efforts to establish a non-arbitrary boundary between the properties of the central controller and the properties of the rest of the agent's body. In one sense it is obvious that the computational demands facing the control system for a body will vary according to the abilities of and the constraints upon that body. Furthermore, it is well known that the physical configuration of, for example, a robot body can reduce the computational burden on its central controller. Rodney Brooks (1991) and Valentino Braitenberg (1986) showed how bodies or machines can overcome apparently complicated behavioral challenges by virtue of their mechanical structure alone without recourse to representations. Most famously, for example, Braitenberg's vehicles are simple robots whose motors and sensors are connected so as to exhibit behaviors (such as tending to move towards or away from a light source) which are goal directed without the need for any explicit central controller or information processor to determine the body's movement.

In the case of an organism with a relatively limited behavioral repertoire, it is easy to imagine an arrangement whereby all evolutionarily relevant computational challenges are resolved via the physical structure of the body itself. The morphology of a bacterium is suited to the set of computational challenges that a bacterium might encounter in its evolutionary niche.

For agents with more complicated sets of behaviors and challenges, it is reasonable to assume the involvement of some kind of central control system. However, drawing the line between the central controller and the body in computational problem solving is difficult. Without carefully determining the capacities of the body, it remains an open question just how much of the computational burden with respect to some problem is being assumed by the central controller. Likewise, the capacities and constraints associated with the body itself can be understood as part of the problem landscape to be navigated by the central controller. In robotics, Rolf Pfeifer drew attention to this set of questions in his discussion of the morphology and control trade-off problem (Pfeifer and Scheier 1999).

The morphology and control problem involves determining an agent's bodily capacities and constraints, but it can also be understood as the problem of determining the boundaries between the computations that are performed by the central controller and those that are performed by the rest of the agent's body.

This problem raises very traditional philosophical questions concerning embodiment. Some advocates of a view known as morphological computation, most notably Chandana Paul, have argued that bodies should be understood as part of the agent's cognitive process

(Paul 2004). Paul defined morphological computations as computations where the mechanical structure of the body itself carries some of the burden of solving the agent's computational problem. The goal of her research is to provide "a common currency between the realm of the physical body and the controller" in robotics (Paul 2006), and she writes:

Controllers lie in the realm of abstract computation, and as such are usually implemented in computational hardware. However, if physical interactions can also perform computation, it becomes possible for the dynamics of the morphology to play a computational role in the system, and in effect subsume part of the role of the controller. (ibid., 620)

So, for example, Rolf Pfeifer describes a robot hand that is designed with flexible and soft gripping surfaces, artificial tendons such that a single instruction from the hand's control system can initiate a range of kinds of gripping actions with a range of different kinds of objects. The key here is that the physical structure of the robot hand carries the burden of coping with a wide range of behavioral challenges. The manner by which the robot hand grips a wide variety of objects could be understood as a highly complex information theoretic challenge. Much of the complexity of that challenge is overcome by means of the morphological structure of the hand.

5.3 Perceptrons, bodies, and computation

Chananda Paul's morphology-based treatment of the Boolean XOR function provides a clear way of understanding the central conceptual problems in morphological computation. Paul (2004) describes a robot which, by virtue of its morphology, can exhibit XOR-constrained behavior. The robot has a central control system composed of perceptrons, but that system, for reasons to be explained below, cannot implement the XOR function. Instead, the rest of the agent's body implements a function that looks a great deal like XOR. As we shall see, it is not quite correct to straightforwardly identify the morphologically computed solution with XOR.

The significance of the XOR function is connected to the history of cognitive science, and specifically to Rosenblatt's development of the perceptron and the subsequent criticisms by Minsky and Papert (1969). A perceptron is a network which takes multiple inputs and gives one output. Rosenblatt describes the perceptron as consisting of three layers: The input layer, the layer of association units, and the output layer. The output of the middle layer is a threshold weighted sum of the inputs. The network has adjustable synaptic weights and fires once it reaches some value determined by its threshold. Rosenblatt proved the perceptron convergence theorem, which states that a perceptron can compute any linearly separable function. Historically, the inability of perceptrons to learn to solve functions which are not linearly separable was one of the reasons that popular scientific opinion turned against neural networks for a time.

A linearly separable problem is one in which for any output neuron there must be some hyperplane (of dimension n-1) which divides the set of n inputs to that neuron between those which activate the output neuron and those that do not. So, if there are two inputs with values of either 0 or 1, given that the problem is linearly separable, there should be a one-dimensional hyperplane (a line) which partitions the set of outputs into a set consisting exclusively of 0s and another consisting exclusively of 1s.

In the case of Boolean operations like AND or OR, there is linear separability. This is easy to see, given a truth table style representation of these operations.

OR		
Input 1	Input 2	Output
1	1	1
1	0	1
0	1	1
0	0	0

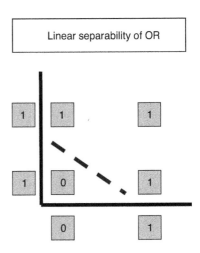

Linear separability of OR

By contrast, functions like XOR or XNOR are not linearly separable as can be seen in the following table and figure:

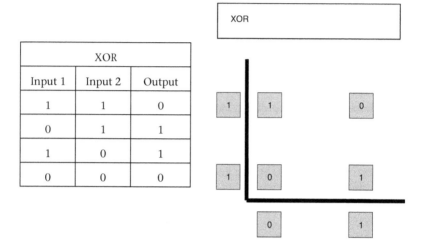

XOR		
Input 1	Input 2	Output
1	1	0
0	1	1
1	0	1
0	0	0

XOR is not linearly separable as can be seen from the figure here. No straight line will partition the space of values into homogenous sets of 0s and 1s. Given that functions like XOR and XNOR are not linearly separable, they fall beyond the computational capacity of a single perceptron.[6]

The robot which Paul describes is controlled by two perceptrons, one for a motor M1 that computes OR, the other for a second motor M2 that computes AND. Whereas M1 turns a single wheel permitting forward motion, M2 serves to lift the wheel off the ground. By means of these two simple perceptron-governed patterns of behavior, the robot is able to perform the XOR function.

A	B	Behavior
T	T	Stationary
T	F	Moving
F	T	Moving
F	F	Stationary

The XOR Robot has one wheel, with two actuated degrees of freedom. The motor M1 is responsible for turning the wheel so that the robot moves forward. The motor M_2 is responsible for lifting the wheel off the ground. Each motor is controlled by a separate perceptron network, which takes as inputs A and B. M1 is controlled by a network which computes A OR B, and M_2 by a network which computes A AND B. Using only these controllers, the robot is able to display the XOR function in its behavior. (See Paul 2004, 34.)

Since an AND perceptron and an OR perceptron by themselves cannot compute the XOR function, her robot relies on the structural features of its body to perform the XOR computation. Specifically, we can understand the structure of the robot as providing a way of generating a function which is computationally equivalent to the standard resolution of the XOR problem with a three-layer feedforward network. The difference here is that it is the body, and not another layer of connectionist processing, which solves the problem. As Paul (2004) points out, the body is not merely an effector but part of the "cognitive" equation itself.

How should we understand the role played by the body in the computational equation? Well, we can expand on her description of the truth table for the various parts of the robot as shown in Table 5.1.

However, we might be concerned with providing an account of the functional role of the morphology of the body itself (call it morph). For, example, we might want to specify the function as shown in Table 5.2.

Table 5.1 Truth values for functions and associated robot actions

A or B	M1	A and B	M2	A xor B	Robot
T	GO!	T	Lift	F	Don't move
T	GO!	F	Don't lift	T	Move
T	GO!	F	Don't lift	T	Move
F	STOP	F	Don't lift	F	Don't move

Table 5.2 Truth table for the function "morph"

(a	or	b)	morph	(a	and	b)
T	T	T	F	T	T	T
T	T	F	T	T	F	F
F	T	T	T	F	F	T
F	F	F	F	F	F	F

Notice that though this specification of morph captures the behavior of the robot under all possible configurations of inputs for the combination of the AND and OR perception, it does not exhaustively characterize morph as a logical operator. Specifically, the table above lists only three of the 2^n combinations of truth values for morph. Filling in the final line in the truth table for the operator morph is pretty straightforward, as can be seen in Figure 5.1.

Paul's description of morphological computation provides a precise example of the kind of embodiment that proponents of the embedded-embodied approach in philosophy of mind envision.

So, how should we interpret Paul's characterization of morphological computations? To begin with, we should be cautious about drawing metaphysical implication from examples like this. The mere existence of morphological computation does not, by itself, present a significant challenge to the view that morphological computations are just as multiply realizable as central computations albeit within the context of an "extended functionalist" framework. Functions realized via morphological computation are fully compatible with traditional multiple realizability. In principle, different configurations of the body could provide different extended ways of implementing *the same* function. In the case of morphological computations, one might imagine a variety of different ways in which the structure of the body could supplement the perceptrons, computationally speaking.

So, the point here is not to present a metaphysical challenge to multiple realizability taken as an in principle argument. Rather, examples like this support a proposal to modify classical constraints on computational theory in cognitive science. To begin with, by fixating on multiple realizability, traditional cognitive science has tended to assume the fixity

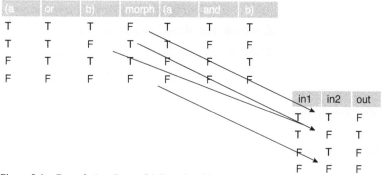

Figure 5.1   Completing "morph's" truth table

of the function at the computational level of description. According to the perspective we'd advocate here, the unique morphological features that the XOR* robot (as we'll call it) exemplifies, are constitutive of the computational challenge facing the agent as well as the behavioral repertoire which it exhibits in response. Computing XOR* morphologically is not the same thing as modeling the approximation of the XOR function with a disembodied network. As the robot agent navigates its environment, the details of its embodiment and the range of feasible physical interactions available to it are constitutive of the manner in which it learns to implement XOR*.

Furthermore, details external to the cognitive system itself can affect the computational problem facing the system; consider how the environment can play a role in the problem, when, for example, we place the XOR* robot on a steep slope. In this case, the function, as described at the computational level, is altered, and as Paul notes such changes happen *as a whole* (Paul 2004). This is why we would call the function XOR* rather than XOR. The XOR function is impervious to hills.

For cognitive science, ignoring bodily and environmental constraints is a methodologically risky strategy insofar as it can miss the problems that serve as the starting point for computational theory. Once these are acknowledged, our computational accounts of the agent's challenges and responses are subject to revision and can be updated in light of increased understanding of the body and the environment.

In this sense, the computational theory of how the agent responds to the environment can be understood as an ongoing research topic, rather than a matter of settled functions at the computational level.

On the view we are advocating here, computational theory should reveal precisely what Marr said it does: It should illuminate *the goal of the computation,* explain *why it's appropriate, and* discover *the logic of the strategy by which it can be carried out.* The computation itself *is* the result of the agent finding or *falling into* a strategy for achieving some goal.

This strategy is constrained by the body, the environment, and the goals of the agent. However, one might object that the morphological computation movement risks confusing the kinds of problems that minds must overcome with the kinds of problems that bodies or organisms as a whole must overcome. The mind (for the most part) solves its own problems in the service of solving the agent's problems.

One problem is the claim that we could think of this as pushing the problem back a bit. If we were, for example to identifying the mind with some aspect of the sub-agential level, we might, for example, push back to identifying the activity of the mind with the functions performed at the level of the perceptrons. If so, then how does the central controller (read the perceptions) tackle the problem of dealing with this bodily morphology? But isn't there something arbitrary about identifying the central controller with the perceptrons rather than any other aspects of the machine? Moreover, there are two distinct perceptrons connected by a body; which of the two is to be identified with the central controller? The AND perception or the OR perceptron? Notice too that they are jointly participating in the effort of producing the XOR operation.

Marr's criticism of the study of feature detecting neurons and his criticism of Gibson certainly encouraged the view that Marr understood the computational level as autonomous with respect to the biology of perceptual systems. However, the *problems and strategies* view of computation which we should take from Marr is more fruitful and, strikingly, not alien from the embodied-embedded. Reflecting on morphological computation encourages the view that computational theory, as Marr understood it, is already the study of embodied computation.

Historically, the bias in favor of the functional or computational level of investigation dates back at least to the classic early works in cognitive psychology and is clearly articulated in one of the most influential early works of that tradition, Ulric Neisser's *Cognitive Psychology* (1967). Clearly, the doctrine of multiple realizability makes plenty of room for the view of the mind as a symbol manipulating system whose relationship to the body is at least contingent and most likely irrelevant. However, our quarrel in this paper is not with multiple realizability per se. Rather, it's with the idea that we have a true and complete story with respect to the way the mind works at the computational level. If we had such a story, then it's quite possible that the body would be irrelevant. However, we will only arrive at a good computational account by understanding the kinds of problems faced by cognitive agents, and we will only understand those problems by understanding the capacities of the body.

The most obvious point we can take from examples like the XOR* robot is that computational theory can generate distinct stories about the inner workings of the agent even when the resulting function is understood. This is not very surprising. Where methodological considerations are more dramatically altered is with respect to the conditions defining a problem for the organism or agent. Understanding what can be morphologically computed by an agent requires more attention to the way the body works than philosophers of mind have been used to paying.

Notes

†We are very grateful to audiences at the University of Pittsburgh and at the New Mexico-Texas Philosophical Society for helpful feedback. Thanks especially to Sarah Robins, whose commentary on an earlier version of this paper helped us to improve it considerably. We gratefully acknowledge the support of Fundación Séneca (Agencia de Ciencia y Tecnología de la Región de Murcia), through project 11944/PHCS/09, and The Ministry of Economic Affairs and Competitiveness of Spain, through Project FFI2009–13416-C02–01.

1. For more on Marr's methodology see Symons 2007.
2. We are grateful to an anonymous reviewer for encouraging us to add this section.
3. They focus on what they see as the mereological fallacy involved in claiming that the brain rather than the whole person "gathers information, anticipates things, interprets the information it receives, arrives at conclusions, etc."
4. The example that Bickle cites is Jerry Fodor's antireductionism. He writes:
   Citing associative processes and the effects of emotion on perception and belief, even the Jerry Fodor of 1975 insisted that explaining them is left to lower-level (probably biological) investigation (1975, 203). Yet Fodor has long denied that reduction is viable for theories about genuinely cognitive phenomena. (Bickle 1998, 3)
5. Some philosophers insist that we have a finished theory of mind, namely our folk psychological theory of mind. We assume that our readers share our hope for more from the science of mind than the belief-desire-action model of mind has provided so far. If you do, then you do not think that the game is over yet and accept that we do not have a finished theory of mind.
6. Although of course it is now possible to construct neural networks which are not subject to this constraint and can solve non-linearly separable problems.

References

Bennett, M. R. & Hacker, P. M. S. (2003). *Philosophical Foundations of Neuroscience*. Oxford: Blackwell.

Bickle, J. (1998). *Psychoneural Reduction: The New Wave*. Cambridge, Mass.: The MIT Press.

Braitenberg, V. (1986). *Vehicles: Experiments in Synthetic Psychology*. Cambridge, Mass.: The MIT press.

Brooks, R. A. (1991). Intelligence without Representation. *Artificial Intelligence*, 47(1), 139–159.

Chalmers, D. J. (1996). *The Conscious Mind: In Search of a Fundamental Theory*. Oxford: Oxford University Press.

Davidson, D. (1970). Mental Events. In *Actions and Events*. Oxford: Clarendon Press.

Davidson, D. (1993). Thinking Causes. In John Heil and Alfred Mele (eds), *Mental Causation*, Oxford: Clarendon Press.

Dummett, M. A. (1993). *The Logical Basis of Metaphysics*. Cambridge, Mass.: Harvard University Press.

Fodor, J. A. (1975). *The Language of Thought*. Cambridge, Mass.: Harvard University Press.

Kripke, S. A. (1980) *Naming and Necessity*. Cambridge, Mass.: Harvard University Press.

Kusch, M. (1995). *Psychologism: The Sociology of Philosophical Knowledge*. London: Routledge.

Marr, D. (1982) *Vision: a Computational Investigation into the Human Representation and Processing of Visual Information*. New York: Henry Holt and Co.

McIntyre, L. (1999). Davidson and Social Scientific Laws. *Synthese*, 120(3), 375–394.

Minsky, M. and Papert, S. (1969). *Perceptrons*. Cambridge, Mass.: The MIT Press.

Neisser, U. (1967). *Cognitive Psychology*. New York: Appleton-Century-Crofts.

Paul, C. (2004). Morphology and Computation. *Proceedings of the International Conference on the Simulation of Adaptive Behavior* (pp. 33–38), Los Angeles.

Paul, C. (2006). Morphological computation: A Basis for the Analysis of Morphology and Control Requirements. *Robotics and Autonomous Systems* 54(8), 619–630.

Pfeifer, R. and Scheier, C. (1999) *Understanding Intelligence*. Cambridge, Mass.: The MIT Press.

Putnam, H. (1975). *Mind, Language and Reality. Philosophical Papers* (Vol. 2). Cambridge: Cambridge University Press.

Symons, J. (2007). The Complexity of Information Processing Tasks in Vision. In Carlos Gershenson, Diederik Aerts, and Bruce Edmonds (eds) *Philosophy and Complexity* (pp. 300–314). Singapore: World Scientific.

6
Embodied Collaboration in Small Groups
Kellie Williamson and John Sutton

6.1 Collaboration, teams, and sport

Being social creatures in a complex world, we do things together. We act jointly. While cooperation, in its broadest sense, can involve merely getting out of each other's way, or refusing to deceive other people, it is also essential to human nature that it involves more active forms of collaboration and coordination (Tomasello 2009; Sterelny 2012). We collaborate with others in many ordinary activities which, though at times similar to those of other animals, take unique and diverse cultural and psychological forms in human beings. But we also work closely and interactively with each other in more peculiar and flexible practices which are in distinctive ways both species-specific and culturally and historically contingent: from team sports to shared labor, from committee work to mass demonstrations, from dancing to reminiscing together about old times.

One such form of collaboration is team sport. Playing with others as part of a team can be a hugely satisfying, stimulating experience. It can also, often simultaneously, be difficult and fraught with moments where collaboration and co-ordination break down. The task demands of team sports are challenging and intricate, often pushing players to the limits of their abilities. Yet even amateur sports teams experience moments, games, or seasons when the team works well together. How does such fast-paced, embodied collaboration operate? Consider the corner kick in football. Corner kicks can be an opportunity for both teams to reset their focus, often after a panicked scramble by the defense to stop an attacking raid. For the attacking team, a corner can be an opportunity to implement a series of moves or a set play of which the team has shared knowledge and practised experience. For the defending team, the pre-occupation is with moving the ball out of the goalmouth and out of danger as quickly as possible. While, descriptively, we can tease out each team's broad aim, the heat of the moment is usually a messier affair. There's typically a flurry of less than calm communication, with defending players, often the goalkeeper, shouting directions, and attackers jostling shoulder to shoulder with defenders vying to be first to the ball. Players must recurrently and effectively switch their focus from their nearby opponents to specific fellow team members and the kicker. Despite this organization and routine, the way a corner kick plays out depends on many small, sometimes game-changing contingencies: how quickly the kick is taken, whether it overshoots the target player or falls short, landing at the feet of a defending player, or whether (on rare but magic occasions) a kicker manages a direct shot on goal that must be defended. The ability to adapt pre-arranged, shared routines to contingencies in the environment lies at the heart of the collective skills involved in a corner kick. In short, the corner kick, a microcosm of the wider complexity found in a football match, demands that teams somehow balance their reliance on pre-prepared routines and courses of action with joint, complementary responses to unexpected, fast changes in the world. This takes a dynamic mix of verbal cues, hand signaling, body positioning, and role distribution, operating alongside and in conjunction with the team's response to unexpected error or unexpected movement by an opponent, when there just isn't time for explicit communication. Given such demands and the unruly contingencies of the sporting environment, team sport is a promising research arena for the study of collaborative and coordinated action.

Though the pervasive nature of collaboration in human life is no big news in the social sciences in general, philosophers of mind and cognitive scientists, with their own traditions and preoccupations, have only recently turned in numbers to theorizing and studying such socio-cognitive interactions, asking what kinds of adaptive and shared intelligence are involved in collaborative activities of many kinds. Recent shifts within philosophy of mind and cognitive science have seen many theorists broaden their questions and practices to focus on complex and intricate cognitive and affective processes that spread beyond a single individual's brain – distributed across the body and/or the environment, coopting objects and driving interactions with other individuals. If we follow Robbins and Aydede (2009) in using the label 'situated cognition' as the broadest descriptive category for these new-wave cognitive theories, we can include under its wing a range of distinctive research traditions, with some common ancestry and assumptions but each increasingly developing its own claims and style: work on embodied cognition (Clark 1997; Chemero 2009; Shapiro 2010), enactivism (Varela, Thompson, & Rosch 1991; Thompson 2007; Di Paolo & De Jaegher 2012), distributed cognition (Hutchins 1995, 2010a; Kirsh 1995; Sutton 2006), and the extended mind hypothesis (Clark 1997; Clark and Chalmers 1998; Rowlands 1999). These frameworks have drawn attention, in different ways, to the role of activity, the body, external objects, and social interaction in cognition, tracking the cognitive significance of the world beyond the skin and skull. While the approach to studying embodied, skillful collaboration in team sport that we outline here is compatible with many of these streams of research, it is not dependent on any of them. We stress three initial methodological and theoretical points.

First, the extended mind hypothesis, for example, has drawn the attention of researchers to the role that aspects of the environment play in individual cognitive processing, including the kind of cognitive scaffolding that language and tools provide. Until recently, however, the emphasis in philosophical work on the extended mind has been on person-object interactions, including the use of pen and paper, notebooks, human-computer interfaces, iPhones, and so on. In contrast, we follow other enthusiasts for *socially* extended cognition (Wilson 2005; Tollefsen 2006; Hutchins 2010b; Sterelny 2010; Gallagher 2013) in seeking to explore interactions between different people, even when artifacts and objects are also part of the setting or cognitive ecology. In particular, we want to draw attention to the interactions themselves, and the processes that drive joint action between people, rather than the storage and retrieval of single mental states from another person. We advocate doing this via theorizing about real world contexts in which complex, intelligent, and skillful collaborative behavior emerges, and drawing on empirical research where data is collected from such contexts, or laboratory methods are adapted to better emulate real world conditions, enhancing what psychologists call 'ecological validity' (Pinder et al. 2011; Barnier 2012).

Second, work in situated cognition and its subspecies does not, as is sometimes thought, discourage or seek to bypass study of the brain. Rather, as is perhaps clearest in the distributed cognition framework and in the 'second-wave' complementarity-based versions of the extended mind hypothesis (Clark 1997; Sutton 2010), they promote a distinctive kind of 'brain theory.' The functioning of neural systems in humans is unusually plastic and open to influence, heavily reliant on social and cultural resources in both development and mature operation, particularly prone to hook up with and adapt to the peculiar characteristics of the artifacts, institutions, and settings which we have collectively constructed (Clark 2003; Hutchins 2011).

Finally, alongside integration with new movements in cultural neuroscience and neuroanthropology (Chiao 2009; Lende & Downey 2012), the various schools of situated and embodied cognition therefore need a more sustained focus on real, culturally embedded, skillful bodily practices. Rather than simply mentioning jazz improvisation, circus, dance, or sport as metaphors of or models for dynamic, densely interactive coupling between embodied cognitive agents, or between groups of agents and their tools, standard empirical cognitive science can fruitfully be integrated both with existing rich resources of music, dance, and sport psychology and the kinds of thick description and close analyses of the microprocesses of interaction which are sought in cognitive ethnography (Goodwin 1994, 2013; Hutchins 1995; Enfield & Levinson 2006; Williams 2006; Sawyer 2007; Streeck, Goodwin, & LeBaron 2011; Geeves et al. 2014). But we retain the ambition of integrating such ethnomethodological detail back into mainstream cognitive theory. The framework we're seeking thus differs clearly both from classical individualist cognitive science, where cognition occurs only in the head of individual thinkers, and from those forms of social theory in which the mind has gone missing entirely, displaced in favor of analyses of ideology or symbolism (as noted also by Downey 2010; Tribble & Sutton 2012). This particular middle ground is far from empty. But in applying these situated or distributed approaches to skilled movement in small groups, we expand its integrative interdisciplinary reach into distinct areas of both philosophy and psychology.

6.1.1   Recent treatments of sport

Singling out one domain of independent interest as an instructive case study, we focus especially on research from sports psychology. Sport is an area of socio-cognitive and bio-cultural endeavor that is surprisingly underexplored in the cognitive sciences and yet is ripe for investigating human cognition. The bulk of existing research on sport highlights individual expertise alone. In his memoir *Sport*, for example, Colin McGinn discusses what it is like to play many diverse individual sports, but mentions team sports only in relation to spectators, tribal sentiments, and hero worship. At the heart of skillful sporting experience, for McGinn, is the individual embodied mind, engaged with "the obliging yet resistant world" or the single "deadly opponent": the passion for sport derives from "experiencing yourself as a unity," and from the "sense of autonomy" gained in "pleasant existential aloneness," the "muscular solipsism" of "a singular being, with my own force and will" (McGinn 2008, 2, 38–40, 117). The fact that people often play sport *with* others, as well as against them, disappears in McGinn's treatment.

The majority of experimental research on skill acquisition in sport has focused on individual sports performance. This research is well developed, with a variety of established and robust experimental paradigms aimed at identifying the psychological characteristics of elite athletes, including expert patterns of perception, attention, and anticipation (Mann, Abernethy, & Farrow 2010; Müller et al. 2010); and fast decision-making and pattern recognition (Gorman, Abernethy, & Farrow 2011). The research is typically conducted with individual athletes performing a subset of tasks that do not require a real contribution from a team member: for instance, returning a tennis serve, hitting a cricket ball or kicking a football at a video-simulation of a match scenario. Creating scientifically robust tasks that capture real features of competition and performance in sport is an ongoing challenge for sports science. Yet, this corpus of research has made significant progress in understanding how individual elite sportspeople can act swiftly and reliably in difficult, changing contexts.

In contrast, our interests in both sport and embodied minds center here on cases in which people are not working or moving alone, in which individual participants' unique skills and capacities are merged and coordinated with those of others in service of shared goals. In such cases, sometimes, the experienced unity of successful flowing performance can perhaps be spread or distributed across many people, when the shared, interactive, and often highly pleasurable sense of autonomy which arises among members of an expert team may be quite unlike the existential aloneness of a singular being. While, as an important first step to understanding how teams perform successfully, it is essential to understand the processes that govern individual performance, we seek to push beyond this, to a fuller picture of team skills.

6.1.2 A tool kit for cognitive collaboration

While in other contexts we defend the idea that groups or pairs of people may themselves form cognitive systems or group minds (Sutton 2008; Williamson & Cox 2014; Theiner, Allen, & Goldstone 2010; Theiner 2013) our focus here is more modest. In constructive rather than critical mode, we sketch elements of a framework for understanding socio-cognitive interactions in teams, taking sports teams as exemplars of collaboration more generally. Collaborative cognition in teams, we argue, is driven by the complex interplay of what, for convenience rather than metapsychological accuracy, we will call higher-level and lower-level processes. Each category, as we use these terms, involves a range of heterogeneous processes that can be usefully grouped together. Higher-level processes are, to a first approximation, the kinds of things that team or group members can talk about: the kinds of processes that can be rendered explicit, as in the use of written or verbal information sharing, or even deliberate, iconic bodily cueing, like pointing or hand waving. These processes can be plans, strategies or instructions made and shared before or after a match, or changed and adapted during play, but they can also include more immediate verbal cues or directions used on the fly to signify an intention or to influence the attention of a team member. In some contexts, they can also include the use of formalized or formalizable game plans, visually represented for instance through diagrams, video footage or on-field/court reenactment. This list is not intended to be exhaustive, but to exemplify the kinds of processes that may be usefully characterized as higher-level.

Lower-level processes are those that are not immediately, easily, or perhaps ever able to be tapped by talk. They include gestural, bodily, and movement-based forms of information-sharing and cueing, often driven by skillful and honed perceptual and attentional processes. These processes are often thought of as implicit and non-deliberative. They can be fast and adaptive, but they are also developed and shaped through practice and performance history. Broadly, lower-level processes are those processes that rely on non-verbal forms of communication and information-sharing: anticipating and responding to the bodily presence of a team member; the direction, speed, and shape of a team member's run; the feel and rhythm of the team's movement.

These inclusive labels do not name single unified mental processes, but rather a heterogeneous mix of processes. Each category is internally diverse and complex, as will be reflected in the range of examples falling within each category. The distinction between higher-level and lower-level processes merely offers an initial conceptual framework for tracking the possible modes of interaction that influence a team's performance. We make this distinction partly because different research traditions focus on only one kind of process at the expense of the other. For example, social ontology in philosophy of mind and organizational psychology focus almost exclusively on higher-level processes, gesturing only in passing at the existence of implicit processes. Similarly, some phenomenological accounts of expertise and skilled performance focus solely on bodily and environmental attunement to the exclusion of higher-level processes (Dreyfus 2007a, 2007b; see Sutton et al. 2011). In contrast, it is the interaction between higher- and lower-level processes that is of intense interest to us, and we stress the dense and complex interplay between them. An appreciation of how such diverse processes interact is essential for understanding how human collaboration is achieved and maintained.

6.2   Our collaborative lives

Instances of collaboration and social interaction are abundant in our daily lives, though the two are not coextensive. Sometimes other people are merely external causal influences on individual cognition: a cruel comment from the crowd, for example, can distract a player at a key moment of the match, disrupting performance. Sometimes, in turn, social topics or the skills and characteristics of other people are merely what individuals think *about*, the content of individual cognitive states: the same player may be motivated to produce a brilliant performance by bringing to mind during the match thoughts or feelings about a particular opponent or mentor. But, we argue, these are not yet cases of collaborative cognition in any substantial sense. Genuinely collaborating, acting, thinking, performing, and reminiscing together are unlike cases of more accidental social interaction on a number of dimensions (Barnier et al. 2008; Sutton 2008). While there are domain- and task-based differences across distinctive sorts of joint actions and cognitions, we can also identify some fundamental properties and processes of collaboration that feature in many kinds of joint action. We focus on sports teams in subsequent sections, but first we sketch other examples of the kind of dense interactions that can emerge between two or more people in typical collaborative, socio-cognitive interactions.

One of the most common forms of cognitive collaboration, perhaps surprisingly, is the activity of remembering. It may feel and seem to us that remembering our past experiences is very much an individual experience, where we not only own our memories, but are the sole influence on how our memories are recalled, what it is we remember and what it is we forget. But a large percentage of our conversations with others involve reminiscing about our past (Dritschel 1991; Bohanek et al. 2009). Our own autobiographical memories, in many cases, may be richly shaped not only by the people with whom we share experiences, or those we encode memories with, but also by the people with whom we share, retrieve, and reconstruct our memories (Pasupathi 2001; McLean & Pasupathi 2011). Some memories, while capturing the gist of what took place, seem to be more malleable when recalled in the context of certain listeners (Campbell 2003, 2008) and can be influenced by social factors like conversational norms (Skowronski & Walker 2004).

An everyday example is a family remembering together. They might share an experience, and then later spontaneously reminisce together about this experience. It might happen, for instance, on the way to visit a grandparent, or to a familiar holiday destination, or maybe over dinner or around the television. In these situations different family members may contribute different aspects of the recalled memory: someone might remember when it was, someone else might remember who was there, and someone else might remember that a poignant anecdote now commonly spoken about first surfaced on this particular holiday (Hirst, Manier, & Apetroaia 1997; Fivush, Bohanek, & Duke 2008; Shore 2009). There are also more systematic ways that couples or families might remember together, involving complex practices such as using a communal calendar that helps regulate who needs to be where and when, or looking through old photographs together (de Frias, Dixon, & Bäckman 2003; Brookfield, Brown, & Reavey 2008; Wu et al. 2008; Harris, Barnier, & Sutton 2013; Bietti & Galiana Castello 2013). Particular family members might also come to be responsible for remembering specific kinds of information such as phone numbers, or the names of distant relations, knowledge of which can be invoked by other family members' requests (Wegner, Giuliano, & Hertel 1985; Wegner 1987). While remembering together informally differs from forms of collaboration which might arise through formally joining a team or work group, we can get a sense from these cases of the complex ways in which members of a collaborative endeavor subtly interact, shaping and influencing each other's cognition.

Though there certainly are embodied shared processes at play in the above cases (indeed, we suspect that many of the lower-level bodily processes we highlight below would be operative in cases of shared remembering), we move now to cases of movement-driven embodied collaboration, where co-actors must respond to and in concert with each other with their bodies and their movements. Just as a sports team must coordinate their actions efficiently and effectively, so too must a company of modern dancers. Working together and with a choreographer both in the development of a dance and then in performance, dancers must develop and act on a similar understanding of the shared task (Stevens et al. 2003; Sutton 2005; Kirsh 2013). In rehearsal and preparation, there is time to share and practice, but the mechanisms by which complex bodily skills and movement sequences can be collectively assessed, maintained, transformed, and honed are hard to sustain and thus to study (Kirsh 2011). In performance, there is often no time or opportunity for explicit verbal communication. And because many decisions and intentions are not easily captured by language, there is more demand for different modes of communication between co-actors, including postural and gestural modes of communication. Dancers working together must be sensitive and responsive to each other's movements, in some way attuned to each other's decisions and actions. Sensitivity to each other's actions is not merely a social task but also a cognitive task which requires attention, memory, and assessments of when and how to act. The sort of effective collaboration required here involves the interplay of various strategies developed in rehearsal with more in-the-moment processes. In many forms of modern dance, the performance is not the same every time. So while actions planned in prior rehearsal are key, so too are more immediate, synchronic processes that allow the dancers to adapt to novelty or error. In later sections, we canvass the kinds of higher- and lower-level processes that interact to sustain embodied, skillful collaborative performance in sport, but we do so in the spirit of using sport as an exemplar of embodied collaboration more generally. We are quietly optimistic that processes similar to those we describe in sporting collaboration may also be efficacious in other kinds of embodied collaboration such as contemporary dance (Sutton 2005).

Working outwards from examples of pairs and families remembering together and from non-sporting cases of embodied collaborative cognition, we round out these examples by specifying further some of the striking aspects of performance in team sports. The demands sports place on players' perception, action, and decision-making capacities are severe, such that successful coordination between team-mates in a team sport is a particularly impressive cognitive feat as well as a sign of the technical and physical prowess of the team members. In team sport players must respond to the often unpredictable behavior of the opposition, and must anticipate the actions of fellow team-mates during an unexpected change in play. Team-mates need to share information about the field of play that cannot be directly accessed by other team members, as when one player has their back to the rest of the field with no time to adjust their position and has no direct visual access to judge the best option for distributing the ball. Similarly, a rugby team must identify the right conditions for a particular set play, and when variations on this set play need to be implemented. Then rapidly and efficiently this decision must be made accessible to other team members.

Some of this play will of course depend on each individual's own skills and expertise, but it will also depend on how the team-mates interact and share information, both in prior practice and at speed in game time, so as to maximize resources. In cases like these, motivation, cognition, and communication must align to such an extent that it is superfluous to distinguish cognitive, communicative, and social processes. What matters here is identifying key processes that facilitate the collaborative completion of a cognitively demanding task. What are the various higher- and lower-level processes involved? And how do these interact?

Our examples in this section are diverse but not exotic or unusual. They bring to the fore questions and phenomena to guide conceptual and empirical inquiry into cognitive and social coordination, or shared cognition. A plausible framework for explaining thinking and acting with others should accommodate such a diversity of cases. While our focus here is on outlining a broad framework for explaining collaboration in team sport, we are not assuming that all sports are the same, or that all teams or athletes are the same. Instead, we seek out a tool kit to help us explain how collaboration is achieved while also accommodating the diversity of sports team performance.

6.3   Theorizing higher-level processes

Many sports involve fast-paced, high-stakes tasks, where athletes must act decisively, precisely and often with little time for careful deliberation, acting on-the-fly and in the moment. In addition, in team sports athletes are required to coordinate their actions, complementing and assisting each other, in the pursuit of a shared goal or aim. It is tempting to theorize sports team behavior solely in terms of implicit, unreportable, non-verbal, or deliberative processes where athletes are guided mostly by fast, lower-level processes governed by their direct, present engagement with and attunement to aspects of the task at hand. While in the next section we highlight some examples of lower-level processes, we draw first on literature that emphasizes the higher-level team processes that also constrain and influence a team's behavior. The kinds of higher-level processes we have in mind are deliberative planning and explicit knowledge-sharing processes. For example, game plans devised off the field by coaching staff or players operate as broad sketches of the shape that the team's performance should take including position allocations and tactical options. Plans of this kind are necessarily general, to allow for the details to be filled in the specific situations that arise through the dynamics of a particular game. Yet these plans can still be extremely subtle, and the subject of intense secrecy, speculation, and analysis. Other forms of higher-level processes might also include specific instructions in the form of verbal feedback from coaches or athletes before, during, or after competition, and in training contexts. Idiosyncratic verbal cues, code-words or metaphors used by one athlete to another in the heat of the moment are a common means of ensuring that two or more players are making compatible, complementary decisions. Sometimes, for instance, in various football codes, these processes may be formalized through diagrammatic representations of game strategies: such diagrams represent 'plays' or 'moves' by depicting the proposed actions of players in specific positions at specific times, often taking into consideration expected responses from players in the opposing team. To provide some theoretical backing for the observation that planning and explicit knowledge sharing play an important role in sports team performance, we can look to organizational psychology and social ontology.

6.3.1   Organizational psychology

A common thread of inquiry in organizational psychology characterizes group behavior in terms of information processing (Hinsz, Vollrath, & Tindale 1997). According to this view, groups process information in much the same way as individual cognizers. The focus in this area of research is on what makes a group or a team of people, often cooperating in formal, organized situations, perform successfully and efficiently. Research in this area identifies a mix of cognitive and social factors that interact to drive group performance, including factors that influence group motivation, and the group's capacities to search, attend to, and process select information (Cannon-Bowers, Salas, & Converse 1993; Klimoski & Mohammed 1994; Eccles & Johnson 2009). Models of this kind have also been applied to sports teams (Reimer, Park, & Hinsz 2006; Eccles & Tennenbaum 2007). Many of these models of team processes highlight the importance of shared knowledge across a team (Cannon-Bowers & Salas 2001; Poizat et al. 2009). The team's ability to compile this shared knowledge over time, access it and update it grounds the capacity for coordinated action: knowing what to do, and when and with whom to do it. We can speculate that the kind of shared knowledge a sports team needs to have might relate to the sport in general, including knowledge of rules, positions, and general success conditions, but also more team-specific knowledge, or shared semantic knowledge of team tactics and various players' skills and tendencies. This notion of differential expertise is helpfully captured by David Eccles' concept of division of labor in teams. Working within sports psychology, Eccles draws on constructs from organizational psychology to highlight the importance of both 'sharedness' and also differentiation within groups (Ward & Eccles 2006). Shared knowledge of the team and the task, he contends, is important for and mediates coordination and collaboration in teams. The knowledge that needs to be shared includes plans, strategies, and meanings of idiosyncratic terms and verbal cues (Eccles & Tenenbaum 2004). As well as shared knowledge, for Eccles, successful groups or teams typically tend to have a decided division of labor between the different group members, often based on heterogeneous expertise in the group (Eccles & Tennenbaum 2007; Eccles & Johnson 2009; Wegner 1987). Importantly, for a team to perform successfully, they must be able to coordinate this expertise and labor effectively (Eccles 2010). On our framework, this may be achieved through a mix of higher- and lower-level processes, for example, both verbal instructions and game plans and more subtle detection of gestural and bodily cues for team members.

While abstract and conceptual models of information processing in groups are useful for a broad-brush account of the high-level processes in team interactions, there are important questions left unanswered. Specifically, to integrate better with cognitive theory we want to investigate the mechanisms responsible for a group searching for and attending to different information, and updating shared knowledge. Organizational psychology enables us to break down the stages involved in collective information processing, but for a richer framework of collaboration we suggest supplementing specific organizational psychology models with conceptual analysis from social ontology, and evidence of the kind of low-level processes of acting and thinking with another.

6.3.2   Social ontology

Another useful set of conceptual resources for identifying and characterizing higher-level team processes can be found in the area of philosophy known as social ontology. The field deals explicitly with questions of what it is for multiple people acting together to share mental states, and what this means for each individual's mentality. This is an informative area of analysis for our framework because it assists us in carving out what is different, in terms of higher-level processes, for people genuinely acting as part of a team or group or pair, when compared to mere aggregates of individuals who only appear to be acting together: think for example of the differences between walking with a friend or group of friends, and walking alongside unknown people on a street

(Gilbert 1990). Where philosophy of mind has a history of considering human mentality as isolated from the social realm (Fodor 1980), social ontology takes sociality and social interaction as its object of philosophical inquiry. We can adapt some of these projects to explore further the cognitive characteristics of people acting together.

For instance, John Searle (1990) argues that the differences between accidental social interaction and genuine collective or collaborative action relate to the intentions involved in the collective action. Genuine collective action, according to Searle, involves collective, shared intentionality, or 'we intentions,' where each participant has an intention to undertake the activity together. Since Searle's early analysis of genuine collaborative activity, theorists have sought to identify the kinds of shared mental states that drive genuine collaboration. Margaret Gilbert (1987, 1992) claims that social interaction involves the sharing of mental states between or across multiple individuals. For Gilbert, people who are committed to acting together accrue a set of (non-moral) duties and obligations, including a duty to see through the shared action. For example, members of a sports team are expected to remain members of the team, playing and training together. Unexpected and unexplained departure from the team is not usually met with approval and acceptance by fellow team members (see also Gréhaigne 2011). Gilbert's account highlights the need for each of the individuals acting together to be aware that they are acting with others, and to be mutually aware of the joint nature of their intentions. For our framework, mutual awareness can be plausibly understood as both a social and cognitive phenomenon, relying on social factors such as the distribution of roles across the people acting together, knowing who is to do what, and communicating this information between those involved, as well as cognitive factors such as attention to and perception of the behavioral cues of one's co-actors, and joint decision-making.

Michael Bratman builds on the social ontology tradition carved out by Searle and Gilbert, emphasizing the interrelatedness of multiple individuals' intentions when acting together. For Bratman, collective or joint action involves an interpersonal structure of connected intentions composed of the intentions of each of the individuals participating in the action (Bratman 1992, 2009). The way in which the individual's intentions are interrelated is through the "meshing of subplans." On Bratman's view, intentions are what guide our planning and preparing for future action. For an action to be joint or collective, and successful, actions must be suitably coordinated. So not only does joint or collective action require each person to have a "we-intention," or an intention that refers to the shared action and the other participant(s), but there must also be compatibility in the way in which the action will be executed, in how the subplans mesh.

While Bratman's focus on the interconnectedness of people who are acting together usefully captures important features of the higher-level processes involved in collaboration, the emphasis on planning and pre-determined courses of action suggests that Bratman's account does not have high-speed collaborative embodied skills as its primary domain of application. Many of our collaborative activities involve spontaneous responses to unpredictable circumstances (Preston 2012), and often coordination is not a matter of planning and verbally communicating those plans. Many team sports, at least in part, require players to inno-vate and act together in the face of an unpredictable, changing task environment. Sure, part of a team's behavior is best explained in terms of plans, strategies, and set plays including, for example, plans regarding when to bring on particular players from the bench, the enacting of a specific attacking set play in the final moments of a game, or specific verbal cues that tell a team-mate what's about to a happen. And, sure, each player has a we-intention to play together. But there will also be times when error or what the referee or the opposition does disrupts the team's pre-planned strategies. In situations like these, the team must rely on other cognitive resources, especially resources that are dynamic and immediate. Explaining this sort of on-the-fly group action in terms of we-intentions and the meshing of subplans does not seem to be the most complete approach. Instead, some sort of combination or interplay between we-intentions and lower levels of information processing may be more likely to drive group or team performance: a dynamic inter-play of higher-level and lower-level processes. These possibilities are explored in the next section. For our broader framework, analysis from social ontology is useful for teasing out the higher-level processes that are present in collaboration. Ideally, analysis of shared mental states, like we-intentions and shared knowledge, could be usefully integrated with research on lower-level team processes.

The notion of shared or collective mental states (shared intentionality) has received a surprising breadth of empirical treatment in the neuro- and cognitive sciences. Some interesting empirical anchors for the kind of shared or collective intentionality analyzed in social ontology have emerged in recent brain-imaging research. This work lends some credence to the possibility that certain higher-level processes or modes of thinking arising in collaboration have distinctive neural underpinnings. Andreas Roepstorff's research program seeks to develop experimental paradigms for brain-imaging "mutually interacting minds." In its early stages, social neuroscience of this kind sees researchers developing new methodologies to try to recreate genuine collaboration in a lab environment with varying degrees of success. Given the manifold of processes that may be operative during collaboration, designing workable imaging experiments is immensely challenging. Konvalinka and Roepstorff (2012), in a recent review, map some of the difficulties with social neuroscience of this kind, including the design of ecologically valid tasks that can be undertaken in accordance with the restrictions on movement in, for example, fMRI scanning. Early findings include a tendency for the brain rhythms or frequencies of two people to synchronize when undertaking a shared task that requires mutual interaction (Dumas et al. 2010) and shifting differences in activation in specific brain regions of interest between senders and receivers of information in turn-taking collaboration (Montague et al. 2002). While these studies have limitations (Konvalinka & Roepstorff 2012), this direction of research suggests a fruitful integration of theoretical approaches to higher-level processes with investigations of neural signatures (Schilbach et al. 2013).

A more established example of empirical investigations into the nature of shared intentionality and we-mode thinking is the fascinating research with human children and non-human primates from Michael Tomasello's lab. Tomasello and his school of researchers work with great apes and infants investigating whether, and if so under what conditions, these primates collaborate by way of shared intentionality or, more specifically, shared goals and intentions. On the basis of a vast breadth of research Tomasello and colleagues (Tomasello et al. 2005; Tomasello 2009) argue that while great apes are able to detect intentionality (mental states) in a fellow ape, they do not do so by way of joint intentions or the sharing of mental states. This body of research suggests that we-intentions or we-mode thinking is unique to our species. Great apes seem able to appreciate that others have mental states, such as goals or intentions, but they appear unable to engage in the sharing of these through culturally complex collaboration (Tomasello et al. 2005). For example, great apes do not seem able to sustain joint attention. Tomasello and colleagues argue that something other than an understanding that a fellow actor has mental states is required for collaboration, namely, shared intentionality. The relationship between these empirical projects and relevant theorizing about the nature of high-level processes or shared intentionality is an ongoing, difficult project, but is all the while making inroads into creating a picture of the cognitive and neural mechanisms that underpin these higher-level process (Gallotti & Frith 2013).

6.4   Finding lower-level processes

One of the most common observations by sports spectators, commentators, and researchers alike, is the speed and intensity at which athletes decide and act in sporting performance – prima facie, in many sporting contexts, there seems to be no time to deliberate on a course of action and then pursue it. As we've suggested, higher-level processes may play a kind of action-constraining or action-guiding role, but it's unlikely they do so without significant facilitation from lower-level processes that are responsive to the specific context in which an athlete or team must act right now. From practitioners' lore, we suspect that gesture or bodily movements and cues are the kinds of things team members are attuned to, and that the information carried by bodies in this way may be used to guide team members' actions. Of course, the specific cues that an athlete attends to may vary depending on the sport: in football, team members may be able to detect information visually from a team member, whereas, in contrast, in team rowing the kinds of information athletes attune to may still be bodily, but may be more likely to be a kind of rhythmic, movement-based information that is felt rather than visually detected (Millar, Oldham, & Renshaw 2012). These kinds of embodied processes have been studied in various disconnected areas of research, but not specifically for sports teams. Empirical research that taps lower-level processes of the kind we're interested in can be very difficult, given that many of these processes operate quickly, automatically and beneath self-report. To get a sense of how existing research might apply to sports teams, we need to be a little inventive in our theorizing. As we noted at the outset, sports science has focused primarily on individual performance, further adding to the need for integrative theorizing. To cash out our understanding of possible lower-level team processes we need to borrow again from research outside sport.

The most compelling evidence of the way two or more people interact on the basis of immediate, bodily, automatic processes comes from a branch of research that we can call 'alignment studies' following philosopher Deborah Tollefsen (Tollefsen & Dale 2012; Tollefsen, Dale, & Paxton 2013). In these studies, the emphasis is on the automatic, unconscious ways in which interacting with others constrains our cognition. 'Alignment' refers to the ways in which two or more people coordinate their cognitive and behavioral states, without necessarily trading in fully reportable mental states like intentions or beliefs. When people act together, they do so by affecting each other's behavior across multiple levels of cognitive processing. Implicit and often unreportable cues such as eye gaze patterns, changes in body movement and rhythm, posture, and gestures play an important role in facilitating and sustaining successful collaboration. As we will see, processes of alignment are useful for beginning to explain fast-paced joint action such as team sports or spontaneous cooperative action. While more research is needed to identify such lower-level processes of alignment in sports performance, there is robust evidence for their existence in cognitive psychology experiments, and as we will see, there are suggestions that this kind of research is possible with sports teams in a real world setting.

As an example of the kind of alignment processes we are interested in, we take a study by the experimental psychologists Richardson and Dale (2005) on the relationship between speakers' and listeners' eye gaze patterns. Richardson and Dale sought to identify to what extent a pair's eye gaze patterns are 'coupled' or constrained by the communication between the two people. Four participants' speech was recorded describing a picture of the cast from a TV series. The remaining thirty-six participants listened to the recordings. Both sets of participants had their eye movements tracked, either while describing the picture or listening to the recorded description, depending on whether one was a speaker or listener. In analyzing the patterns of movement, Richardson and Dale found that the pattern of the listeners' eye movements matched the pattern of the speakers' eye movements, with a two-second delay. They further found that the more similar the listener's eye movements were to the speaker's, the better the listener's comprehension of the speaker's speech, when questioned about the content. The significance of this is that the verbal cues provided by one person direct the gaze of the other, constraining what information they attend to. As the speaker was not present when subjects listened to the description, this is good evidence that the verbalizations alone influenced the listener's attention, not gesture or posture. The verbalizations are not just carrying content about the picture but are also playing a role in constraining what the listener attends to. Typically, gaze patterns are not something of which we are consciously aware, so the striking thing about this study is that the actions and cognitions of another person affect one's own cognitive processing beneath conscious awareness. This kind of evidence helps us to build on key concepts from organizational psychology and social ontology by providing evidence of lower-level, more automatic ways in which two people can coordinate their behavior. Where mutual awareness and we-intentions are likely to be reportable, processes of alignment are beneath the level of report, and yet still efficacious in the completion of a shared task. This study is not only striking for the lower-level processes it reveals, but also for the interplay between higher- and lower-level processes that it suggests. The implicit and dynamic process of eye-gaze movement to detect key information in the task environment is constrained and guided by the verbal behavior of a co-actor. On our characterization of higher- and lower-level processes, this is an intermeshing of high-level processes adopted by one co-actor, namely verbal descriptions, and lower-level processes adopted automatically and unconsciously by the other co-actor.

In a study more relevant to movement, Richardson and colleagues (2007) examined the presence of alignment in rocking chair movements. This is more sports-relevant because it deals with co-actors' perception and action mutually influencing each other. The researchers investigated whether interpersonal coordination would occur between two people, either intentionally or unintentionally, when sitting side-by-side in rocking chairs. In the first experiment, investigating intentional coordination, 24 participants in 12 random pairs were instructed to coordinate their movement. There were two conditions whereby pairs were instructed either to fix their gaze directly ahead, seeing their partner only peripherally, or on the arm rest of their partner's chair. Participants were told to coordinate their rocking either with gaze fixed ahead or on the other's arm rest, depending on which condition they were part of. Surprisingly, Richardson and colleagues found that there was no difference in the stability of coordination between both conditions. This suggests that when coordination is intentional, the information needed to achieve it can be picked up either focally or peripherally.

In order to create as close an analogue to real world social encounters as possible, Richardson et al. performed the same experiment again, but without instructing the pairs to coordinate their actions. In experiment 2, 16 new participants unknown to each other were assigned randomly to pairs. So as to maintain the coordination's status as unintentional, participants were told that they were testing the ergonomics of the chair and must be tested together to save time. As with the first study there was a focal gaze condition and a peripheral gaze condition, disguised to the participants as a test of postures in the chairs. Participants wore earmuffs to block out auditory cues. Participants in each condition were asked to practice rocking, and they were instructed to ignore their co-participant. It was found that unintentional rocking occurred for those participants who were visually coupled, with participants focusing on each other's arm rest. Coordination, or the synchronizing of their movements, was achieved when the movements of the chairs correlated better than chance. This study highlights two important things. The first is that two people can coordinate their actions automatically and without conscious intent. The second is that this is done on the basis of detecting visual information about the others' movements. This is further evidence of lower-level processes by which two or more people can shape each other's actions and cognitions, coordinating automatically and swiftly.

We draw now on research in sports psychology as evidence of the kinds of visual information team members are able to detect in perceptually difficult circumstances (Williams & Ericsson 2005). Kylie Steel and colleagues (Steel, Adams, & Canning 2006) provide evidence to suggest that athletes' perception is attuned to identifying familiar athletes based on minimal information. Working with 15 touch football players, Steel presented the players with 400msec video clips of people running, whose familiarity to the participants varied from high to low. The footage was created using point-light displays, which meant that most of the distinctive information about the runner was omitted, leaving only the mechanics of their gait as represented by the movement of the point-lights. The runners were a mix of participants' team-mates, players from opposition teams and unfamiliar non-touch football players. Participants were asked to identify anyone who was familiar, and to rate their certainty. Strikingly, participants were significantly above chance at recognizing familiar runners, including team-mates and opponent players. This was despite both the short duration of the clips, and the substantial reduction of distinct information. Steel and colleagues replicated this study and its findings with water polo players and their swim stroke (Steel, Adams, & Canning 2007). These findings are very interesting from a collaborative perspective, as they suggest the possibility that team members, over time (though how much time it's not clear), become attuned to specific information from other team members. By combining the evidence from alignment studies that people attune their bodily movement to each other when acting together with Steel's work, we get a sense of the way that team members use immediate information from other team members (e.g. information about gait or movement style), detected quickly and near automatically (like alignment processes), to collaborate. These are the kinds of low-level processes and mechanisms that contribute importantly to collaboration.

Moving to more fully collaborative sport contexts, we can again look to evidence from sports psychology to provide a hint as to how automatic, lower-level cognitive processes interact with participants' mental states, like we-intentions and intermeshed subplans, and with shared knowledge: essentially, the dense interaction between higher- and lower-level processes. With this evidence we can see that embodied collaboration involves a complex interplay of a pair's or a group's mental states with moment-to-moment processes. In research conducted with doubles tennis teams, Blickensderfer and colleagues (2010) investigated the link between players' previous experience, their shared expectations, and the implicit co-ordination between team members. The construct of 'shared expectations' refers to one domain or sub-set of shared knowledge. Where shared knowledge can span a number of domains including knowledge about the task and about specific team members, long-standing background knowledge, or knowledge that is specific to a given situation (Poizat et al. 2009), shared expectations involve knowledge of or team agreement about expected responses from the team. Blickensderfer et al. claim that both shared knowledge and shared expectations are acquired with experience. They predicted that the more experience a team had, the more similar or shared their expectations would be regarding the team's response to match situations. The researchers made the further prediction that the more shared the team's expectations were, the more the team would rely on implicit (non-verbal) coordination rather than explicit, verbal coordination. The predicted relationships between these constructs provide a nice test case to suggest how higher-level processes might interact with and shape lower-level processes, namely that forms of shared knowledge can mediate the use of non-verbal communication. Worth noting here though, is that unlike the alignment cases above where the participants were strangers, the participants in this current study have a history of playing together – they have had time to build and refine their shared knowledge.

The researchers surveyed 71 high-performing teams to ascertain the degree of their task familiarity (years/time spent being coached), team familiarity (years/time playing together), and the extent or degree to which team members shared expectations about team responses to match situations. Shared expectations were measured by players being separately shown drawings of tennis doubles scenes, offered several possible team responses and asked to rate the likelihood that their own team would select each response. The teams' matches were then filmed. Raters of the footage identified the degree of "relative position" between team members, as a measure of implicit co-ordination. Relative position refers to team-mates "adjust(ing) and adapt(ing) their positioning with respect to each other's positioning during team performance" (Blickensderfer et al. 2010). Relative position is akin to the kind of intelligent but fast bodily or positional movement in our football examples, and could equally apply to dancers' movements or to a mother and child re-enacting or recollecting a past experience. Relative position is achieved under severe time constraint and need not rely on overt communication between team members. This is not unlike many mundane cases of collaboration where the action may not be easily verbally communicated due either to time pressures or the difficulty in describing the task linguistically – for example, high-pressure work environments like industrial kitchens or surgical wards. As with some of the alignment evidence, the presence of relative position in this study suggests that more than shared knowledge or we-intentions are needed to explain collaboration.

When the survey and video ratings were correlated, Blickensderfer et al. found that the degree to which a team had shared expectations about responses to match situations predicted the amount of relative positioning (implicit coordination) that the team uses. This suggests that there is a link between higher-level processes like shared knowledge and lower-level processes like bodily communication or information sharing. This study reflects what we have been describing as the interplay between higher- and lower-level processes: in this example, the team's shared knowledge, arising through their history of playing together and available for explicit report, mediates the use of communication processes, enabling a range of implicit, non-verbal, bodily forms of communication. Implicit coordination here is in some ways akin to what we have been calling lower-level processes. But where our characterization of lower-level processes differs is that implicit processes are not *all* mediated by shared knowledge – as we saw with some of the alignment examples, these processes can also operate in the absence of shared knowledge or at least with very minimal shared knowledge, given that alignment can operate among strangers.

This study is a nice example of the possibility that higher-level processes interact with and are mutually shaped by more immediate, lower-level processes. Research of this kind helps to identify the variety of ways in which information can be transmitted amongst team members, and how this information shapes athletes' attention and decision-making. In situations where environmental conditions are changing and unpredictable, such as when faced with a fast-paced opposition, alternate ways of sharing information between team-mates, especially through more implicit channels and cues such as body positioning, gestures, and movement, may be vital.

6.5   Conclusion

We have characterized collaboration as involving complex and hetero-geneous higher- and lower-level processes, which are both usually in operation and interaction. In doing this we explored the ways in which different forms of embodied collaboration draw on different kinds of processes, verbal and non-verbal, planned and innovative, to sustain collaboration in tough, cognitively challenging task environments. Drawing on previously disconnected fields of research, we have sought to bring more integrative balance to existing research on collaboration that has often emphasized one kind of process at the expense of the other, thus encouraging a productive engagement between these contributions to our theoretical toolkit. Higher- and lower-level processes can be and should be teased out analytically, but their interconnectedness is what we think really drives embodied interactions between team and group members.

References

Barnier, A. J. (2012). Memory, Ecological Validity and a Barking Dog. *Memory Studies*, 5(4), 351–359.

Barnier, A. J., Sutton, J., Harris, C. B., & Wilson, R. A. (2008). A Conceptual and Empirical Framework for the Social Distribution of Cognition: the Case of Memory. *Cognitive Systems Research*, 9(1), 33–51.

Bietti, L. M. & Galiana Castello, F. (2013). Embodied Reminders in Family Interactions: Multimodal Collaboration in Remembering Activities. *Discourse Studies*, 15(5), 665–686.

Blickensderfer, E. L., Reynolds, R., Salas, E., & Cannon-Bowers, J. A. (2010). Shared Expectations and Implicit Coordination in Tennis Doubles Teams. *Journal of Applied Sport Psychology*, 22(4), 486–499.

Bohanek, J. G., Fivush, R., Zaman, W., Lepore, C. E., Merchant, S., & Duke, M. P. (2009). Narrative Interaction in Family Dinnertime Conversations. *Merrill-Palmer Quarterly*, 55(4), 488–515.

Bratman, M. E. (1992). Shared Cooperative Activity. *The Philosophical Review*, 101(4), 327–341.

Bratman, M. E. (2009). Modest Sociality and the Distinctiveness of Intention, *Philosophical Studies*, 144(1), 149–165.

Brookfield, H., Brown, S.D., & Reavey, P. (2008). Vicarious and Post-Memory Practices in Adopting Families: the Construction of the Past in Photography and Narrative. *Journal of Community and Applied Social Psychology*, 18(5), 474–491.

Campbell, S. (2003). *Relational Remembering: Rethinking the Memory Wars*. Lanham: Rowman & Littlefield.

Campbell, S. (2008). The Second Voice. *Memory Studies*, 1(1), 41–48.

Cannon-Bowers, J. A. & Salas, E. (2001). Reflections on Shared Cognition. *Journal of Organizational Behavior*, 22, 195–202.

Cannon-Bowers, J. A., Salas, E., & Converse, S. (1993). Shared Mental Models in Expert Team Decision-Making. in N. J. Castellan, Jr. (Ed.), *Individual and Group Decision Making*, 221–246. Hillsdale, NJ: Erlbaum.

Chemero, A. (2009). *Radical Embodied Cognitive Science*. Cambridge, MA: MIT Press.

Chiao, J. Y. (2009). Cultural Neuroscience: A Once and Future Discipline. *Progress in Brain Research*, 178, 287–304.

Clark, A. (1997). *Being There: Putting Brain, Body, and World Together Again*. Cambridge, MA: MIT Press.

Clark, A. (2003). *Natural-Born Cyborgs: Why Minds and Technologies Are Made to Merge*. Oxford: Oxford University Press.

Clark, A. & Chalmers, D. (1998). The Extended Mind. *Analysis*, 58(1), 7–19.

de Frias, C.M., Dixon, R.A., & Bäckman, L. (2003). Older Adults' Use of Memory Compensation Strategies Is Related to Psychosocial and Health Indicators. *Journal of Gerontology: Psychological Sciences*, 58, 12–22.

Di Paolo, E. & De Jaegher, H. (2012). The Interactive Brain Hypothesis. *Frontiers in Human Neuroscience*, 6, 163.

Downey, G. (2010). "Practice without Theory": A Neuroanthropological Perspective on Embodied Learning. *Journal of the Royal Anthropological Institute*, 16(s1), S22–S40.

Dreyfus, H. L. (2007a). Response to McDowell. *Inquiry*, 50(4), 371–377.

Dreyfus, H. L. (2007b). The Return of the Myth of the Mental. *Inquiry*, 50(4), 352–365.

Dritschel, B. H. (1991). Autobiographical Memory in Natural Discourse. *Applied Cognitive Psychology*, 5(4), 319–330.

Dumas, G., Nadel, J., Soussignan, R., Martinerie, J., & Garnero, L. (2010). Inter-Brain Synchronization during Social Interaction. *PloS One*, 5(8), e12166.

Eccles, D. (2010). The Coordination of Labour in Sports Teams. *International Review of Sport and Exercise Psychology*, 3(2), 154–170.

Eccles, D. W. & Johnson, M. B. (2009). Letting the Social and Cognitive Merge. In S. D. Mellalieu & S. Hanton (eds). *Advances in Applied Sports Psychology* (pp. 281–316). London: Routledge.

Eccles, D. W. & Tenenbaum, G. (2004) Why An Expert Team Is More Than a Team of Experts: A Socio-Cognitive Conceptualization of Team Coordination and Communication in Sport. *Journal of Sport and Exercise Psychology*, 26, 542–560.

Eccles, D. W. & Tenenbaum, G. (2007). a Social Cognitive Perspective on Team Functioning in Sport. *Handbook of Sport Psychology*, 3, 264–286.

Enfield, N. & Levinson, S. C. (eds) (2006). *Roots of Human Sociality: Culture, Cognition, and Interaction*. Oxford: Berg.

Fivush, R., Bohanek, J. G., & Duke, M. (2008). The Intergenerational Self: Subjective Perspective and Family History. In F. Sani (ed.). *Individual and Collective Self-Continuity*, 131–143. Mahwah, NJ: Erlbaum.

Fodor, J. (1980). Methodological Solipsism Considered as a Research Strategy in Cognitive Psychology. *Behavioral and Brain Sciences*, 3(1), 63–73.

Gallagher, S. (2013). The Socially Extended Mind. *Cognitive Systems Research*, 25/26, 4–12.

Gallotti, M. & Frith, C. D. (2013). Social Cognition in the We-Mode. *Trends in Cognitive Sciences*, 14(4), 160–165.

Geeves, A., McIlwain, D. J., Sutton, J., & Christensen, W. (2014). To Think or Not to Think: the Apparent Paradox of Expert Skill in Music Performance. *Educational Philosophy and Theory*, 1–18, doi 10.1080/00131857.2013.779214.

Gilbert, M. (1987). Modelling Collective Belief. *Synthese*, 73(1), 185–204.

Gilbert, M. (1990). Walking Together: a Paradigmatic Social Phenomenon. *Midwest Studies in Philosophy*, 15(1), 1–14.

Gilbert, M. (1992). *On Social Facts*. New Jersey: Princeton University Press.

Gorman, A. D., Abernethy, B., & Farrow, D. (2011). Investigating the Anticipatory Nature of Pattern Perception in Sport. *Memory & Cognition*, 39(5), 894–901.

Goodwin, C. (1994). Professional Vision. *American Anthropologist*, 96(3), 606–633.

Goodwin, C. (2013). The Co-operative, Transformative Organization of Human Action and Knowledge. *Journal of Pragmatics*, 46(1), 8–23.

Gréhaigne, J. F. (2011). Jean-Paul Sartre and Team Dynamics in Collective Sport. *Sport, Ethics and Philosophy*, 5(1), 34–45.

Harris, C. B., Barnier, A. J., & Sutton, J. (2014). Couples as Socially Distributed Cognitive Systems: Remembering in Everyday Social and Material Contexts. *Memory Studies*, 7(3).

Hinsz, V. B., Vollrath, D. A., & Tindale, R. S. (1997). The Emerging Conceptualisation of Groups as Information Processors. *Psychological Bulletin*, 121(1), 43–64.

Hirst, W., Manier, D., & Apetroaia, I. (1997). The Social Construction of the Remembered Self: Family Recounting. In J. Snodgrass & R. Thompson (eds). *The Self Across Psychology* (pp. 163–188). New York: New York Academy of Sciences.

Hutchins, E. (1995). *Cognition in the Wild*. Cambridge, MA: The MIT Press.

Hutchins, E. (2010a). Cognitive Ecology. *Topics in Cognitive Science*, 2(4), 705–715.

Hutchins, E. (2010b) Imagining the Cognitive Life of Things. In L. Malafouris & C. Renfrew (eds). *The Cognitive Life of Things* (pp. 91–101). Cambridge: McDonald Institute for Archaeological Research.

Hutchins, E. (2011). Enculturating the Supersized Mind. *Philosophical Studies*, 152(3), 437–446.

Kirsh, D. (1995). The Intelligent Use of Space. *Artificial Intelligence*, 73(1), 31–68.

Kirsh, D. (2011). Creative Cognition in Choreography. In *Proceedings of the 2nd International Conference on Computational Creativity*, 141–146.

Kirsh, D. (2013). Embodied Cognition and the Magical Future of Interaction Design. *ACM Transactions on Human Computer Interaction*, 20(1), 34pp.

Klimoski, R. & Mohammed, S. (1994). Team Mental Model: Construct or Metaphor? *Journal of Management*, 20, 403–437.

Konvalinka, I. & Roepstorff, A. (2012). The Two-Brain Approach: How Can Mutually Interacting Brains Teach Us Something about Social Interaction? *Frontiers in Human Neuroscience*, 6, 215.

Lende, D. H. & Downey, G. (2012). *The Encultured Brain: An Introduction to Neuroanthropology*. Cambridge, MA: MIT Press.

Mann, D. L., Abernethy, B., & Farrow, D. (2010). Action Specificity Increases Anticipatory Performance and the Expert Adavantage in Natural Interceptive Tasks. *Acta Psychologica*, 135(1), 17–23.

McGinn, C. (2008). *Sport*. Stocksfield: Acumen Press.

McLean, K. C. & Pasupathi, M. (2011). Old, New, Borrowed, Blue? the Emergence and Retention of Personal Meaning in Autobiographical Storytelling. *Journal of Personality*, 79(1), 135–164.

Millar, S., Oldham, A. R., & Renshaw, I. (2012). Interpersonal Coupling in Rowing: the Mediating Role of the Environment. *Journal of Sport and Exercise Psychology*, 34(Supp), 110–111.

Montague, P. R., Berns, G. S., Cohen, J. D., McClure, S. M., Pagnoni, G., Dhamala, M., Wiest, M. C., Karpov, I., King, R. D., Apple, N., & Fisher, R. E. (2002). Hyperscanning: Simultaneous fMRI during Linked Social Interactions. *Neuroimage*, 16, 1159–1164.

Müller, S., Abernethy, B., Eid, M., McBean, R., & Rose, M. (2010). Expertise and the Spatio-Temporal Characteristics of Anticipatory Information Pick-Up from Complex Movement Patterns. *Perception*, 39(6), 745–760.

Pasupathi, M. (2001). The Social Construction of the Personal Past and Its Implications for Adult Development. *Psychological Bulletin*, 127, 651–672.

Pinder, R. A., Davids, K. W., Renshaw, I., & Araújo, D. (2011). Representative Learning Design and Functionality of Research and Practice in Sport. *Journal of Sport and Exercise Psychology*, 33(1), 146–155.

Poizat, G., Bourbousson, J., Saury, J., & Sève, C. (2009). Analysis of Contextual Information Sharing during Table Tennis Matches: An Empirical Study on Coordination in Sports. *International Journal of Sport & Exercise Psychology*, 7, 465–487.

Preston, B. (2012) *A Philosophy of Material Culture: Action, Function and Mind*. New York: Routledge.

Reimer, T. R., Park, E. S., & Hinsz, V. B. (2006). Shared and Coordinated Cognition in Competitive and Dynamic Task Environments: An Information-Processing Perspective for Team Sports. *International Journal of Sport and Exercise Psychology*, 4(4), 376–400.

Richardson, D. C. & Dale, R. (2005). Looking to Understand: the Coupling between Speakers' and Listeners' Eye Movements and Its Relationship to Discourse Comprehension. *Cognitive Science*, 29(6), 1045–1060.

Richardson, M. J., Marsh, K. L., Isenhower, R. W., Goodman, J. R. L., & Schmidt, R. C. (2007). Rocking Together: Dynamics of Intentional and Unintentional Interpersonal Coordination. *Human Movement Science*, 26, 867–891.

Robbins, P. & Aydede, M. (eds) (2009). *The Cambridge Handbook of Situated Cognition*. Cambridge: Cambridge University Press.

Rowlands, M. (1999). *The Body in Mind: Understanding Cognitive Processes*. Cambridge: Cambridge University Press.

Sawyer, K. (2007). *Group Genius: The Creative Power of Collaboration*. New York: Basic Books.

Schilbach, L., Timmermans, B., Reddy, V., Costall, A., Bente, G., Schlicht, T., & Vogeley, K. (2013). Toward a Second-Person Neuroscience. *Behavioral and Brain Sciences*, 36(4), 393–414.

Searle, J. R. (1990). Collective Intentions and Actions. In P. Cohen, J. Morgan, & M. E. Pollack (eds). *Intentions in Communication* (pp. 401–416). Cambridge, MA: The MIT Press.

Shapiro, L. (2010). *Embodied Cognition*. Oxford: Routledge.

Shore, B. (2009). Making Time for Family: Schemas for Long-Term Family Memory. *Social Indicators Research*, 93, 95–103.

Skowronski, J. J. & Walker, W. R. (2004). How Describing Autobiographical Events Can Affect Autobiographical Memories. *Social Cognition*, 22(5), 555–590.

Steel, K. A., Adams, R. D., & Canning, C. G. (2006). Identifying Runners as Football Teammates from 400msec Video Clips. *Perceptual and Motor Skills*, 103, 901–911.

Steel, K. A., Adams, R. D., & Canning, C. G. (2007) Identifying Swimmers as Water-Polo Or Swim Team-Mates from Visual Displays of Less Than One Second. *Journal of Sports Sciences*, 25(11), 1251–1258.

Sterelny, K. (2010). Minds: Extended or Scaffolded? *Phenomenology and the Cognitive Sciences*, 9(4), 465–481.

Sterelny, K. (2012). *The Evolved Apprentice*. Cambridge, MA: MIT Press.

Stevens, C., Malloch, S., McKechnie, S., & Steven, N. (2003). Choreographic Cognition: the Time-Course and Phenomenology of Creating a Dance. *Pragmatics & Cognition*, 11(2), 297–326.

Streeck, J., Goodwin, C., & LeBaron, C. (2011). *Embodied Interaction: Language and Body in the Material World*. Cambridge: Cambridge University Press.

Sutton, J. (2005). Moving and Thinking Together in Dance. In R. Grove, C. Stevens, & S. McKechnie (eds). *Thinking in Four Dimensions: Creativity and Cognition in Contemporary Dance* (pp. 50–56). Melbourne: Melbourne University Press.

Sutton, J. (2006). Distributed Cognition: Domains and Dimensions. *Pragmatics & Cognition*, 14(2), 235–247.

Sutton, J. (2008). Between Individual and Collective Memory: Coordination, Interaction, Distribution. *Social Research: An International Quarterly*, 75(1), 23–48.

Sutton, J. (2010). Exograms and Interdisciplinarity: History, the Extended Mind, and the Civilizing Process. In R. Menary (ed.). *The Extended Mind*, 189–225. Cambridge, MA: MIT Press.

Sutton, J., McIlwain, D., Christensen, W., & Geeves, A. (2011) Applying Intelligence to the Reflexes: Embodied Skills and Habits between Dreyfus and Descartes. *Journal of the British Society for Phenomenology*, 42(1), 78–103.

Theiner, G. (2013). Transactive Memory Systems: a Mechanistic Analysis of Emergent Group Memory. *Review of Philosophy and Psychology*, 4(1), 65–89.

Theiner, G., Allen, C., & Goldstone, R. L. (2010). Recognizing Group Cognition. *Cognitive Systems Research*, 11(4), 378–395.

Thompson, E. (2007). *Mind in Life: Biology, Phenomenology, and the Sciences of Mind*. Cambridge, MA: Belknap Press.

Tollefsen, D. P. (2006). From Extended Mind to Collective Mind. *Cognitive Systems Research*, 7(2), 140–150.

Tollefsen, D. & Dale, R. (2012). Naturalizing Joint Action: a Process-Based Approach. *Philosophical Psychology*, 25(3), 385–407.

Tollefsen, D. P., Dale, R., & Paxton, A. (2013). Alignment, Transactive Memory, and Collective Cognitive Systems. *Review of Philosophy and Psychology*, 4(1), 49–64.

Tomasello, M. (2009). *Why We Cooperate*. Cambridge: Cambridge University Press.

Tomasello, M., Carpenter, M., Call, J., Behne, T., & Moll, H. (2005). Understanding and Sharing Intentions: the Origins of Cultural Cognition. *Behavioral and Brain Sciences*, 28(5), 675–690.

Tribble, E. B. & Sutton, J. (2012). Minds in and Out of Time: Memory, Embodied Skill, Anachronism, and Performance. *Textual Practice*, 26(4), 587–607.

Varela, F. J., Thompson, E. T., & Rosch, E. (1991). *The Embodied Mind: Cognitive Science and Human Experience*. Cambridge, MA: MIT Press.

Ward, P. & Eccles, D. W. (2006). Commentary on "Team Cognition and Expert Team: Emerging Insights into Performance for Exceptional Teams." *International Journal of Sport and Exercise Psychology*, 4, 463–483.

Wegner, D. M. (1987). Transactive Memory: a Contemporary Analysis of the Group Mind. In B. Mullen & G. R. Goethals (eds). *Theories of Group Behavior*, 185–208. New York: Springer-Verlag.

Wegner, D. M., Giuliano, T., & Hertel, P. (1985). Cognitive Interdependence in Close Relationships. In W. J. Ickes (ed.). *Compatible and Incompatible Relationships*, 253–276. New York: Springer-Verlag.

Williams, R. F. (2006). Using Cognitive Ethnography to Study Instruction. In *Proceedings of the 7th International Conference on Learning Sciences*, 838–844. International Society of the Learning Sciences.

Williams, M. A. & Ericsson, K. A. (2005). Perceptual-Cognitive Expertise in Sport: Some Considerations when Applying the Expert Performance Approach. *Human Movement Science*, 24, 283–307.

Williamson, K. & Cox, R. (2014). Distributed Cognition in Sports Teams: Explaining Successful and Expert Performance. *Educational Philosophy and Theory* (in press): 1–15, doi 10.108000131857.2013.779214.

Wilson, R. A. (2005). Collective Memory, Group Minds, and the Extended Mind Thesis. *Cognitive Processing*, 6, 227–236.

Wu, M., Birnholtz, J., Richards, B., Baecker, R., & Massimi, M. (2008). Collaborating to Remember: A Distributed Cognition Account of Families Coping with Memory Impairments. *Proceedings of ACM CHI 2008 Conference on Human Factors in Computer Systems*, 825–834.

7
Little-e Eliminativism in Mainstream Cellular and Molecular Neuroscience: Tensions for Neuro-Normativity

John Bickle

Ah, the 1980s: *Roseanne*, *The Cosby Show*, "Let's Dance," U2, Ronnie and Nancy, Princess Di – and eliminative materialism dominating the philosophy of mind (ahem!). Paul and Patricia Churchland didn't invent the view, although they certainly brought it to wide philosophical attention as a central component of their neurophilosophy. At stake was supposed to be the status of our commonsense conception of the mental, our belief-desire "folk" psychology, which grounded not only perennial philosophy, but also our legal institutions and many of our social-interactive practices. Many philosophers took public stands on eliminative materialism – a few for it, most against. Few denied the gravity of what seemed to be at issue.

Yet by the early 1990s interest in eliminative materialism had waned. There were lots of reasons why, though not one of them was because the view was subjected to decisive criticism.[1] By then the dispute had grown mostly "ontological," distancing it from the earlier concerns with new developments in neuroscience and landing it instead into arcane disputes about reference, natural kinds, concepts – all those topics that excite analytical philosophers to no end, but which bore everybody else to tears (neuroscientists included). Even the Churchlands succumbed. Paul's (1981) essay took us "beyond folk psychology," to inspired speculations about social changes a neuroscience-inspired kinematics for cognition might produce. His (1985) essay told us we could learn to introspect our own brain processes as these were being revealed by neuroscientific discoveries. More than half of Patricia's (1986) book described

134

state-of-the-art neuroscience at the time of its writing. Yet by the end of the 1980s the Churchlands' commitments to "extreme scientific realism" were carrying their eliminativist arguments. Their scientific realism at bottom was just as "philosophical" – in the pejorative sense of the term – as the commitments of any armchair metaphysician.[2] The Churchlands' eliminative materialism began as a view committed to understanding how brain science was developing explanations of cognition; it wound up, in the hands of both prominent eliminativists and critics, as yet more metaphysical crinkum-crankum. No wonder scientifically-minded philosophers soon turned to the hot new topic of consciousness, and eliminative materialism quickly became as passé in philosophy as Bruce Willis as David Addison, Jr., in episodes of *Moonlighting*.

And that's too bad. It's my contention here that early on the eliminativists were onto a really important point about *neuroscientific practice* – a point that unfortunately got lost in the ontological rancor that followed. What's more, developments over the past two decades central to mainstream neuroscience have strengthened the case for this point. I'll call this view about neuroscientific practice "eliminativism-with-a-little-e," to distinguish it from Eliminativism (with a big-E), the ontological thesis. It is also my contention that eliminativism-with-a-little-e causes serious tension for the trendy new disciplines of neuro-normativity: neuroethics, neuropolitics, neuroaesthetics, neuro-lit crit, and the like. This is because there's still an unfilled gap between the neuroscience these new fields typically look to in developing their accounts, and what cellular and molecular neuroscience is increasingly revealing to be the actual mechanisms driving behaviors indicative of cognitive functions. Easy linkages across these fields still elude, and that's a problem for neuro-normativity.

7.1 Molecular and Cellular Cognition (MCC): a case study

The neuroscientific field of "molecular and cellular cognition," or MCC, began in earnest more than two decades ago, with the application of genetic-engineering techniques from molecular biology to mammalian nervous tissue. These techniques enabled experimenters to manipulate any cloned gene in living, behaving mammals. Initially entire genes were "knocked out" at the embryonic stem cell stage – literally excised from the entire genome – but soon far more targeted genetic manipulations became available, including conditional knock-outs and transgenic insertions to overexpress specific genes. Now viral vectors are used routinely to limit overexpressed gene activity to both specific cells (neurons) and developmental stages.[3] Experimental techniques standard for three decades in molecular biology generally insure that the targeted proteins in various intra- and inter-cellular signaling pathways are affected as predicted. A host of behavioral protocols long recognized as indicative of specific cognitive functions by animal experimental psychologists are employed in an effort literally to link mind to molecular pathways directly on the lab bench.

Perhaps the best way to introduce the experimental techniques and promise of this field is through a detailed case study. The particular case study I'm about to relate also illustrates quite nicely the "little-e eliminativism" potentially lurking in the emerging discoveries of this field.[4]

Animal learning psychologists have long known that rodent behavior in many long-term memory tests displays the well-known "Ebbinghaus spacing effect." This effect was first discovered and explored more than a century ago in human psychology by the German psychologist who came to lend it his name.[5] Animals (humans included) learn many tasks best when their training episodes are spaced across several learning sessions, with longer non-training intervals separating the training sessions. This is compared to learning rates for animals who learn the task either in one massed training session, or in the same number of training sessions as the first group but with shorter duration non-learning intervals between the training sessions. In each case in controlled experimental settings, animals spend the same amount of time encountering the training stimuli; only the number of overall training intervals or the non-training time between training intervals differs across the groups. Often it is only the learning ratios of animals in the *longer-interval distributed training* group that are statistically significantly greater than those in either the *massed* or *briefer-interval distributed training* groups. Psychologists have explored many details about the most effective training sessions for numerous memory tasks: optimal number of training intervals, length of training intervals, time between training intervals, and the like. But what are the neural mechanisms for the Ebbinghaus spacing effect? Answers to that question have proven elusive.

One cognitive-level mechanistic explanation for this effect has received some acceptance. Spear and Riccio (1994) argue that, with frequent repetitive training – massed or briefer-interval distributed – information acquired on the previous trial is still being processed when new input arrives, disrupting the ongoing processing. Longer delay intervals between training sessions are required to complete full input integration to each stimulus presentation. This cognitive-level explanation is obviously limited in scope: why does learning and memory require "complete integration" of each stimulus presentation? Why can't later stages of the memory integration process function on previously-acquired information while earlier stages handle the new inputs? This cognitive-level story doesn't answer those crucial questions. This cognitive-level explanation also seems *ad hoc* – motivated solely to capture Ebbinghaus spacing data, but with no obvious generalization capacities or explanatory motivations beyond capturing these specific data. So as Genoux and colleagues insist, this and other "time-dependent constraints on learning and memory are poorly understood" (2002, 971).

Recent work from Isabelle Mansuy's lab provides strong evidence that the direct mechanisms for this psychological effect are *molecular* (Genoux *et al.* 2002). During the training episodes, specific neurons recruited into the memory trace for the training stimuli undergo significant synaptic plasticity, inducing late long-term potentiation (L-LTP). Intraneuronally, and with statistically significantly higher probability, neurons that get recruited into the memory traces have increased phosphorylated cyclic adenosine monophosphate responsive-element binding protein (pCREB) activity at the time of training (Han *et al.* 2007, Han *et al.* 2009, Zhou *et al.* 2009). pCREB is an activity-dependent transcriptional enhancer for genes coding for both regulatory and structural proteins. (Transcriptional enhancers bind to sites in the control region of genes to turn on the process of gene transcription.) The end result of pCREB activation in these recruited neurons is the locking of additional "hidden" AMPA receptors into active sites in the post-synaptic density. These additional active receptors make these neurons much more likely to respond later to similar pre-synaptic activity (glutamate release) induced by later presentations of the training stimuli. However, the high neuronal activity required to induce this pCREB-driven synaptic potentiation also activates protein phosphatase 1 (PP1) in these same neurons. Activated PP1 removes the phosphate group from pCREB molecules in the neuron's nucleus. This shuts down their gene transcriptional activities and ultimately the protein synthesis that potentiates the synapses recruited into the memory trace (Genoux *et al.* 2002). Because of the increased frequency of training stimuli presentations during massed and briefer-interval distributed training, action potential frequency in the neurons recruited into the memory traces is high enough not only to activate pCREB to induce L-LTP but also to shut down the activity-dependent pCREB activity directly by way of activity-dependent activated PP1. The end result in these neurons is little L-LTP, and ultimately little learning to be remembered later. However, with longer-interval distributed training, the extended non-training periods between training sessions are long enough to decrease activation frequency in the neurons recruited into the memory traces, to still above the level required to induce L-LTP via pCREB but below the level required to activate PP1. Lessened PP1 activity allows pCREB in those neurons to induce lasting synaptic potentiation, and ultimately produces the learning reliably measured upon later presentations of the training stimuli. The resulting behavioral dynamics are exactly the Ebbinghaus spacing effect (Silva and Josselyn 2002).

Many philosophers and cognitive scientists will find the experimental details of Genoux *et al.* (2002) daunting, but I must present some of these because the arguments for little-e eliminativism in the next section depend critically upon them. I beg readers' patience to follow along. Genoux and colleagues began by engineering a transgenic mutation in a mouse model for a protein one step further upstream in the pCREB-PP1 pathway. Inhibitor-1 (I1) deactivates PP1, blocking its inhibition of pCREB activation. Genoux and colleagues built a transgenic mutant mouse, the I1*, with an additional copy of the *I1* gene combined with a promoter region for α-calmodulin kinase II (αCaMKII). This promoter region limited transgene expression almost exclusively to forebrain regions, including hippocampus. They also used a reversible tetracycline-controlled transactivator (rTA) system so that the transgene was only expressed when the mutant animals were dosed with doxycycline (dox). Standard molecular-biological controls confirmed that the I1* protein was only overexpressed in hippocampus and cortical neurons when the mutants were dosed with dox.

Genoux and colleagues operationalized the Ebbinghaus spacing effect using a mouse object recognition task, a non-spatial memory task known to be hippocampus-dependent. They divided groups of I1* mutants and non-mutated littermate controls into massed training (one 25-minute unbroken training session with each training object), briefer-interval distributed training (five 5-minute training sessions with each training object, with 5-minute non-training intervals separating the sessions), and longer-interval distributed training groups (five 5-minute training sessions, with 15-minute non-training intervals separating the sessions). During test sessions they measured the amount of time animals explored a pair of objects, one object from the training sessions and a novel object. They calculated discrimination ratios by dividing the total amount of time the animal spent exploring the novel object by the total amount of time spent exploring both objects. The standard protocol for the rodent object recognition memory task is that discrimination ratios greater than 0.5 indicate hippocampus-based memory for the training object (based on rodents' natural preference for novelty). Larger discrimination ratios indicate stronger memory for the object already encountered during training.

As expected, Genoux and colleagues' non-mutated controls displayed the typical Ebbinghaus spacing effect at all memory durations, with only those undergoing longer-interval distributed training displaying statistically significantly greater memory for the training objects. The I1* mutants on dox, however, showed statistically significantly greater memory for the training objects for both longer-interval distributed training *and* briefer-interval distributed training, with no statistically significant differences in discrimination ratios between those two groups, and no differences between I1* mutants on dox on briefer-interval distributed training and control non-mutants on longer-interval distributed training. A follow-up study showed that the number of pCREB-activated hippocampus and cortical neurons in the non-mutated control group only increased significantly with longer-interval distributed training, while the increase in pCREB-activated neurons in I1* mutants on dox increased to statistically comparable levels in both the longer-interval distributed *and* briefer-interval distributed training groups. Integrated with prior studies on the role of pCREB in learning and memory, the results of this pair of *Positive Manipulation* and *Non-Intervention* experiments (see Silva, Landreth, and Bickle 2013) provided strong evidence for the pCREB-PP1 interaction causal mechanistic explanation of the Ebbinghaus spacing effect.

The Genoux *et al.* (2002) case study is an excellent illustrative example of the kinds of results and explanations that have been forthcoming for two decades now from MCC labs. It provides a clear example of what Eric Kandel, James Schwartz, and Thomas Jessel meant when they wrote, more than a decade ago in the introductory chapter of their standard textbook,

> This book... describes how neural science is attempting to link molecules to mind—how proteins responsible for the activities of individual nerve cells are related to the complexities of neural processes. Today it is possible to link the molecular dynamics of individual nerve cells to representations of perceptual and motor acts in the brain and to relate those internal mechanisms to observable behavior. (2000, 3–4)[6]

While philosophers and cognitive scientists (including some cognitive neuroscientists) may puzzle over how to skip so many "levels" separating mind from molecular processes, many neuroscientists are doing exactly that, in their ongoing laboratory research. MCC has been one of the fields that has led this endeavor. The case study in this section was but one of at least 100 examples of MCC experiments that would have illustrated just as well its practices, results, and explanatory promise. (For additional case studies, see Silva, Landreth, and Bickle 2013.)

7.2   Little-e eliminativism

Big-E Eliminativism, the ontological thesis, insisted that there are no such things as beliefs, desires, and the rest of the posits of folk psychology. The structure of representations in the brain and the dynamics of their interactions with perceptual systems, motor systems, and each other, slowly being pieced together by neuroscience, differ radically from the propositional attitudes and logic-like computations over their contents. These differences make the possibility of meaningful cross-theory identities remote. Coupled with neuroscience's expected increases in explanatory power, scope, and mesh with the rest of science, these features are the benchmarks of ontologically eliminative cross-theory relations in science's history, akin to the fate that befell caloric fluid thermodynamics, phlogiston chemistry, and the alchemic essences. The Eliminativist predicts that continuing neuroscience developments will distance its empirically better explanations of human behavior even further from folk psychology's. Folk psychology will increasingly be seen as a radically false theory, whose ontology of mind will be replaced, rather than identified with or incorporated into neuroscience's ontology.

Little-e eliminativism, by contrast, is a much more limited claim about what takes place in actual science. Its scope is further limited to sciences pursuing causal-mechanistic explanations. Concerning cognitive functions specifically, little-e eliminativism characterizes science as (1) operationalizing the phenomenon to be causally-mechanistically explained in terms of some pattern of observable behaviors exhibited in some laboratory protocol. Typically this operationalization will use some behavioral protocol widely accepted as indicating the occurrence of the cognitive function in question. This is why protocols like the Morris water maze, contextual conditioning, and the object recognition task employed by Genoux *et al.* (2002) are used so widely in MCC research. There is broad consensus on the cognitive phenomena they indicate and they have been tested for every potential confound anyone has reasonably thought to limit their validity. (2) Experiments of three different types are then run to investigate hypothesized causal mechanistic explanations for the behaviors measured. *Negative manipulations* decrease the probability or intensity of the hypothesized cause and measure its behavioral effects. *Positive manipulations* increase the probability or intensity of the hypothesized cause and measure its behavioral effects. *Non-Intervention experiments* directly manipulate neither hypothesized cause nor effect, but measure their co-occurrence in experimental settings as close to the organism's natural environment and within biologically realistic parameters to find strengths and directions of correlations. Ideally all three of these experiments will have been run for a specific causal hypothesis since each type of experiment makes important contributions to, but also has important limitations for, confirming a causal hypothesis. Results from collections of these experiments for a specific causal hypothesis are integrated in a variety of characteristic ways.[7]

Little-e eliminativism further claims (3) that concepts descriptive of higher levels of biological organization are increasingly eschewed in the causal hypotheses under serious, sustained neuroscientific investigation, especially in fields like MCC. Instead, it is increasingly lower-level kinds, their dynamics, and their organization into pathways and interactions that get investigated experimentally. Finally, and most characteristically, (4) the differences in the structure and dynamics of the lower-level causal mechanisms, as compared to the causal-mechanistic accounts offered at the higher levels, preclude any easy cross-level linkages. Because the lower-level mechanistic explanations fail to map onto the higher level account, the higher level account can no longer be recognized as a correct explanation of the behavior, even as a correct story abstracted away from the lower-level details. The higher-level kinds are thus *increasingly eliminated* from accepted causal-mechanistic explanations of the behaviors indicative of cognitive functions. Bear in mind: This is an assertion about actual scientific practice in fields like MCC – not about ontology!

All four of little-e eliminativism's characteristic features are nicely illustrated by our case study. (1) The Ebbinghaus spacing effect is operationalized for the purpose of MCC experimentation as performance during the testing phase on the rodent object recognition memory task. The protocol is widely accepted as a hippocampus-dependent non-spatial memory task. The number of training sessions with each stimulus object, or the length of the non-training delay intervals separating each training session, operationalize the concepts of 'massed training,' 'briefer-interval distributed training,' and 'longer-interval distributed training.' All this is in keeping with much previous work by animal experimental psychologists employing this common laboratory protocol.

(2) Working with engineered I1* mutants and non-mutated littermate controls, Genoux *et al.* (2002) performed both Positive Manipulation and Non-Intervention experiments to test the causal hypothesis: Longer interval distributed training → Increased pCREB activity & decreased PP1 inhibitory activity in neurons recruited into a specific memory trace → Increased learning and memory.[8] The Positive Manipulation left pCREB activity unchanged while dampening PP1 activity via the pharmacologically-induced overexpression of the *I1** transgene. Mutants displayed the Ebbinghaus behavioral pattern with briefer-interval distributed training, statistically identical to learning achieved by the non-mutated control animals only with longer-interval distributed training. A Non-Intervention Experiment demonstrated that the number of pCREB-active cells in hippocampus and cortex of the mutants overexpressing the I1* transgene during briefer-interval distributed training matched the number reached in the control animals only with longer-interval distributed training. These results integrate smoothly with prior experimental results linking pCREB activity in particular neurons to memory performance in a variety of tasks and to those neurons' recruitment into specific memory traces.

(3) The prevailing cognitive level explanation of the Ebbinghaus spacing effect explained it as the interruption of ongoing processing of recent stimulus presentations due to the faster re-presentations of training stimuli during massed and briefer-interval distributed training. Longer delay intervals between training sessions allowed full input integration to each stimulus presentation. The pCREB activity-PP1 inactivity causal-mechanistic account supported by Genoux and colleagues' experiments makes no use of such higher-level concepts as 'ongoing processing of stimulus presentations' or 'full input integration.' Instead it appeals to activity-dependent intraneuronal activation of pCREB and PP1, their kinase-phosphatase interactions, effects of activated pCREB on subsequent gene transcription and protein synthesis, the induction of L-LTP in specific neurons recruited into the memory traces for specific stimuli, and ultimately the operationalized Ebbinghaus spacing effect. It integrates this hypothesized mechanism with already-established activities in inter-neuronal pathways from hippocampal and cortical neurons out to the neuro-muscular junctions at the motor peripheries. The higher level concepts from cognitive science are no part of these cellular-molecular causal-mechanistic explanations. The cognitive effects have been operationalized behaviorally for the purpose of experimental investigation; the mechanisms explaining the behavior patterns are hypothesized and investigated experimentally at increasingly lower levels of biological organization, namely, intra- and inter-cellular molecular signaling

pathways. Higher level cognitive (and cognitive-neuroscientific!) kinds are eschewed from MCC causal-mechanistic neuroscientific investigations and explanations.

Finally, (4) the considered cognitive explanation of the Ebbinghaus spacing effect is revealed as *false*. The accepted cognitivist account spoke of interruptions of *ongoing* processing of the training information due to too-frequent re-presentations in massed and briefer-interval distributed training regimens (without saying much about how that processing took place or why later stages weren't encapsulated from the effects of too-frequent training stimulus re-presentations). The molecular explanation based on the Genoux *et al.* (2002) experiments, about the interactions between an activity-dependent transcriptional enhancer and its activity-dependent inhibiting phosphatase in individual hippocampus and cortical neurons recruited into specific memory traces, is not a lower-level realization of that basic cognitivist story. In the cases of massed and briefer-interval distributed training, the neuronal activity that activates both transcriptional enhancer and phosphatase *prohibits the initial processing initiated by the training stimulus from even getting started*. The activated pCREB molecules are quickly dephosphorylated by the activated PP1 molecules before the initial stages of gene transcription and protein synthesis that leads to L-LTP, and hence to consolidated learning and memory, get underway. The underlying molecular dynamics show that the hypothesized cognitivist account is false, or at least in need of significant revision, even as a higher-level approximation of the molecular details that cause the behavioral differences that operationalize the Ebbinghaus spacing effect.

Together, features (1)–(4) make MCC explanations like the dueling pCREB-PP1 activities little-e eliminative of accepted cognitive-level explanations. Not every successful MCC discovery of a molecular mechanism for a given cognitive function meets all four conditions – especially (4). Sometimes the discovered detailed molecular mechanisms do mimic some accepted cognitive-level explanation, with the latter then seen as good high-level approximation of the lower-level mechanisms. But some MCC cases meet all four conditions. Extensive little-e eliminativism of accepted cognitive-level explanations is not merely a logical possibility of ongoing work in MCC. Cases demonstrating all four features have already been found.

7.3   Tensions for neuro-normativity

The potential for widespread little-e eliminativism in MCC brings a new tension to much trendy work in neuro-normativity (neuroethics, neuropolitics, neuroeconomics, neuroaesthetics, etc.). That is because these disciplines live mostly off the resources of cognitive neuroscience and its affiliated methodologies: functional neuroimaging, human neurology and psychopathology, electroencephalography, and the like. Accepted mechanistic explanations of cognitive phenomena in all these areas are exactly the potential targets of MCC little-e eliminativism. To the extent that such little-e eliminativist outcomes come forth, neuro-normativity risks building its resources on kinds that won't be linkable with discovered molecular mechanisms of cognitive functions.

Roskies (2002) usefully distinguishes two separate categories of inquiry in neuroethics, one being the "ethics of neuroscience" and the other being the "neuroscience of ethics." The first concerns ethical dilemmas raised by our increasing abilities to image the functioning brain and manipulate its function with drugs, surgical interventions, and other techniques. Autonomy, privacy and other stock ethical concerns are the principal concerns here. The second has to do with what neuroscience is teaching us about the brain bases of our ethical behaviors and attitudes. It's this second neuro-normative endeavor, the neuroscience of ethics (politics, economic behavior, aesthetic judgment, and the like), that is potentially in tension with MCC little-e eliminativism.

Consider a single, prominent example. Based on a variety of behavioral, psychopathological, and functional neuroimaging data, *dual process theories of moral judgment* have become prominent in the neuroscience of ethics (Greene 2009). According to this approach, humans respond to moral dilemmas with two distinct and sometimes competing systems of moral judgment-making: a consequentialist "controlled cognition" system involving prominently dorsolateral prefrontal cortex, and an intuitive emotional response system yielding "deontological" judgments, running through the amygdala, superior temporal sulcus, perhaps the temporal parietal junction, and medial prefrontal cortex. Moral conflict across these systems is monitored by the anterior cingulate cortex. In dilemmas that fail to generate a "prepotent emotional response," such as pulling a switch to divert the trolley and saving the five but killing the one, humans typically default to the consequentialist controlled cognition system and judge accordingly ("it is morally permissible [maybe even obligatory] to pull the switch"). However, in dilemmas that elicit a "prepotent emotional response," such as having to personally push Fatty off the footbridge to stop the trolley to save the five, the intuitive emotional system is engaged, yielding the deontological judgment ("it is wrong to push him to his death"). Dual aspect theorists cite evidence, based mainly on judgments to brief narrative descriptions while subjects undergo fMRI, that the "personal" nature of the harm required in footbridge-type cases, i.e., your actually having to push Fatty off the bridge to his death, is what engages the intuitive emotional system.

Dual aspect explanations of moral judgment take explicit causal-mechanistic form. Consider Greene's own wording:

> We hypothesized that people tend to disapprove of the action in the footbridge dilemma *because* the harmful action in that case, unlike the action in the switch case, *elicits* a prepotent negative emotional response that *inclines* people toward disapproval.... We hypothesized further that people tend to approve of the action in the switch case *because*, in the absence of a prepotent countervailing emotional response, they *default* to a utilitarian mode of reasoning that favors trading one life for five. ... We proposed ... there is an emotional appraisal process ... that *distinguishes* personal dilemmas like the footbridge case from impersonal dilemmas like the switch case. (Greene 2009, 991)

See also his accompanying diagram in Figure 68.2 (2009, 990). There seems little doubt that the arrows in that diagram are most straightforwardly interpreted as causal arrows.

This causal-mechanistic flavor is what makes dual process theories of moral judgment so reminiscent of the cognitivist explanation of the Ebbinghaus spacing effect, little-e eliminated by the molecular pCREB-PP1 interaction account. We're nowhere near yet having a cellular-, much less a molecular-mechanistic explanation of moral judgment. Not one of the four features of little-e eliminativism sketched in the previous section is even under active exploration yet. But the mismatches along those four specific features for cognitive phenomena we now can address experimentally at the level of molecular mechanism suggests that the exclusive appeal to cognitive-level mechanistic explanations, even cognitive-neuroscientific ones, render accounts like the dual aspect theory of moral judgment at least susceptible to a little-e eliminativism looming in their future.

One exception to this tension is current neuroeconomics. Part of that discipline's tradition goes back to primate single-cell recordings of midbrain dopaminergic neurons in decision situations about rewarded choices (Schultz 1998) and micro-stimulation of sensory neurons in situations where decisions about ambiguous visual stimuli involved rewarded choices (Salzman et al. 1992). Because such studies seek correlations (or even mechanisms) directly between cellular activity and decision-making behavior, we have reason to expect that these resources will link rather smoothly with future discoveries of molecular mechanisms generating the neuronal behavior being recorded or manipulated. However, even in neuroeconomics there is a second trend, grounded squarely in cognitive neuroscience – in results garnered strictly by functional neuroimaging, transcranial magnetic stimulation, and the like (Glimcher et al. 2008). Resources informing this second trend also face the potential tension of little-e eliminativism sketched here, as the direct experimental quest for the molecular mechanisms of cognitive functions proceeds. Just because we know that a particular brain region is involved in some behavior, and just because activity in that region has also been correlated with other kinds of behaviors, tells us nothing about *how* activities in the neurons in that region are bringing about the effects. Speaking causally-mechanistically exclusively on the basis of such results makes such accounts potentially susceptible to the little-e eliminativism sketched here.

I'm not suggesting that current gaps between neuro-normative accounts and molecular neuroscience won't be bridged, and in some cases of actual scientific practice bridged relatively smoothly. I am suggesting, though, that the groundbreaking neuroscientific work that is current MCC poses a new challenge for any approach to normativity that claims to be "neutrally plausible." Neurons, including the ones increasingly implicated in the mechanisms of cognitive functions, are nothing more than membrane-bound collections of molecular tricks for getting ions selectively across those membranes. Increasingly those particular tricks are being linked on the MCC lab bench to behaviors we routinely take to be indicators of cognitive functions. This much "mainstream" neuroscience can no longer safely be ignored by anyone engaged in an endeavor that legitimately deserves 'neuro-' in its name. A part of recognizing this fact about current scientific practice involves recognizing the potential for little-e eliminativism about higher-level neuroscientific mechanisms at the behest of developing MCC.

Notes

1. Some still think that the "self-refutation" challenge was decisive against eliminative materialism: that in order to assert a hypothesis, one must believe that hypothesis, so eliminative materialism, which asserts that there are no beliefs, is self-refuting. This challenge was implicit in the many fatuous jokes often tossed eliminativists' way: eliminativists believe there are no beliefs, desire to convince us that there are no desires or convincings, intend by their arguments to show that there are no intentions, etc. (etc., etc.). Yet the

commonplace recognition that when one's available linguistic resources force one to use the very framework one is challenging, one's challenge will appear logically circular until a new "incommensurable" framework is in place, was a point Paul Feyerabend had urged two decades earlier, and illustrated with numerous (non-mentalistic) examples from the history of science. (See many of the essays in his (1981).) Following Feyerabend's lead explicitly, Patricia Churchland once quipped that eliminative materialists "gronkify" beliefs, "where 'gronkification' is a neuropsychological state defined within the mature new theory" (Churchland 1986, 397). In short, the "self-refutation" challenge rests on the assumption that folk psychology must be true because *at the current state of our scientific knowledge of human cognition*, criticizing it requires using it – hardly a knock-down counterargument.

2. Some might balk at my claim here. Didn't both Churchlands, but especially Paul, put resources from "connectionist" neural networks front and center in their later eliminativist arguments? (See especially Paul's (1987) and the later essays in Part I of his (1989).) Even here, their scientific realist ontology lurked closely behind these scientific appeals. Without "extreme" scientific realism, eliminativist conclusions don't follow simply from pointing out the alternative forms of representation and computation at work in connectionist networks.

3. We explain some of the landmark experimental techniques and results from the twenty-year history of this field, through state-of-the-art current research, throughout Silva, Landreth, and Bickle 2013.

4. I've used this same case study to illustrate a number of themes in recent philosophy of neuroscience, including the mechanism-reductionism debate (Bickle and Hardcastle 2013) and some features of the reductionism implicit in current neuroscientific practice that resemble John Kemeny and Paul Oppenheim's (1956) model of scientific reductionism generally (Thereur and Bickle 2013). The Kemeny-Oppenheim model was quickly rejected by philosophers of science as too weak – perhaps too hastily?

5. The scientific content of the next few paragraphs is based on Genoux *et al.* (2002), who used transgenic mice and an object recognition memory task for rodents. For additional details about the actual experiments presented for non-specialists, see Thereur and Bickle 2013.

6. Note that this quote was from a standard mainstream textbook in the field from more than one decade ago. Since then a Fifth edition has appeared in print (Kandel *et al.* 2012), with an even greater emphasis on molecular mechanisms of cognitive functions.

7. All of these concepts, illustrated by numerous examples from landmark MCC experiments, are detailed in Silva, Landreth, and Bickle 2013.

8. The arrows here are the standard causal arrows of causal graph theory. For an overview with applications to MCC experiments, and with references to the primary literature, see Silva, Landreth, and Bickle 2013, chapter 7.

References

Bickle, J. and V. Hardcastle (2012). Philosophy of Neuroscience. *Elsevier Life Sciences Reviews (els)*. DOI: 10.1002/9780470015902.a0024144

Churchland, P.M. (1981). Folk Psychology and the Propositional Attitudes. *Journal of Philosophy*, 78(2), 67–90.

Churchland, P.M. (1985). Reduction, Qualia, and the Direct Introspection of Brain States. *Journal of Philosophy*, 82, 8–28.

Churchland, P.M. (1987). *Matter and Consciousness*, revised edn. Cambridge, Mass.: MIT Press.

Churchland, P.M. (1989). *A Neurocomputational Perspective*. Cambridge, Mass.: MIT Press.

Churchland, P.S. (1986). *Neurophilosophy*. Cambridge, Mass.: MIT Press.

Feyerabend, P.K. (1981). *Philosophical Papers*, vols I and II. Cambridge: Cambridge University Press.

Genoux, D.,Haditsch, U., Knobloch, M., Michalon, A. Storm, D., and Mansuy, I. M. (2002). Protein Phosphatase 1 Is a Molecular Constraint on Learning and Memory. *Nature*, 418(6901), 970–975.

Glimcher P.W., E. Fehr, C. Camerer, and R.A. Poldrack (eds) (2008). *Neuroeconomics: Decision Making and the Brain*, 1st edn. San Diego: Academic Press.

Greene, J.D. (2009). The Cognitive Neuroscience of Moral Judgment. In M.S. Gazzaniga (ed.), *The Cognitive Neurosciences IV*, 987–999. Cambridge, Mass.: MIT Press.

Han, J.-H., S.A. Kushner, A.P. Yiu, C.J. Cole, A. Matynia, R.A. Brown, R.L. Neve, J.F. Guzowski, A.J. Silva, and S.A. Josselyn (2007). Neuronal Competition and Selection during Memory Formation. *Science*, 316(5823), 457–460.

Han, J.-H., S.A. Kushner, A.P. Yiu, H.-L. Hsiang, T. Buch, A. Waisman, B. Bontempi, R.L. Neve, P.W. Frankland, and S.A. Josselyn (2009). Selective Erasure of a Fear Memory. *Science*, 323(5920), 1492–1496.

Kandel, E.R., J.H. Schwartz, and T.M. Jessell (eds) (2000). *Principles of Neural Science*, 4th edn. New York: McGraw Hill.

Kandel, E.R., J.H. Schwartz, T.M. Jessell, S.A. Siegelbaum, and A.J, Hudpeth (eds) (2012). *Principles of Neural Science*, 5th edn. New York: McGraw Hill.

Kemeny, J. and P. Oppenheim (1956). On Reduction. *Philosophical Studies*, 7, 6–19.

Roskies, A.L. (2002). Neuroethics for the New Millennium. *Neuron*, 35(1), 21–23.

Salzman, C.D., C.M. Murasugi, K.H. Britten, and W.T. Newsome (1992). Microstimulation in Visual Area mt: Effects on Direction Discrimination Performance. *Journal of Neuroscience* 12, 2331–2355.

Schultz, W. (1998). Predictive Reward Signal of Dopamine Neurons. *Journal of Neurophysiology*, 80(1), 1–27.

Silva, A.J. and S.A. Josselyn (2002). The Molecules of Forgetfulness. *Nature*, 418(6901), 929–930.

Silva, A.J., A. Landreth, and J. Bickle (2013). *Engineering the Next Revolution in Neuroscience*. New York: Oxford University Press.

Spear, N.E. and D.C. Riccio (1994). *Memory: Phenomena and Principles*. Neeedham Heights, MA: Allyen and Bacon.

Thereur, K. and J. Bickle (2013). What's Old Is New Again: Kemeny-Oppenheim Reduction in Current Molecular Neuroscience. *Philosophia Scientiae*, 17(2) (special issue on "The mind-body problem in cognitive neuroscience"), 89–113.

Zhou, Y., J. Won, M.G. Karlsson, M. Zhou, T. Rogerson, J. Balaji, R. Neve, P. Poirazi, and A.J. Silva (2009). creb Regulates the Excitability and the Allocation of Memory to Subsets of Neurons in the Amygdala. *Nature Neuroscience*, 12, 1438–1443.

8

Ethics and the Brains of Psychopaths: The Significance of Psychopathy for Our Ethical and Legal Theories

William Hirstein and Katrina Sifferd

8.1 Introduction

The first well-publicized success of the neuroscience revolution was Damasio's (1995) case history of a man referred to as EVR, who began to show signs of psychopathy following surgery to remove a brain tumor above the orbits of his eyes. Since then, the neuroscience of psychopathy and sociopathy has steadily moved forward to begin to identify what is different about their brains. Not all psychopathy is caused by environmental conditions, however. There is mounting evidence that some psychopaths are born that way. The existence of such people within a society has profound implications for our attempts to build ethical and just communities. How should we manage the psychopaths amongst us? In this chapter, we will examine this and other questions. We will begin with a description of the current method of diagnosing psychopaths. Then we will describe four competing neuropsychological theories of what is different about their brain functions. In the final sections, we will trace the implications that the existence of psychopaths has for our theories of ethics and of legal and moral responsibility.

8.2 Diagnosing psychopaths

In the early 1800s, doctors who worked with mental patients noticed that some of them who appeared outwardly normal had what they called "moral depravity" (Rush 1812) or "moral insanity" (Pritchard 1835) in that they seemed to possess no sense of morality or of the rights of other people. The term "psychopathy" was first applied to these people around 1900. The term was changed to "sociopath" in the 1930s to emphasize the damage they do to society. Most contemporary researchers have returned to using the term "psychopath." Some of them use that term to refer to a more serious disorder, linked to genetic traits, producing more dangerous individuals, while continuing to use "sociopath" to refer to less dangerous people who are believed to be more products of their environment, including their upbringing (Partridge 1930). Other researchers make a distinction between "primary psychopaths," who are thought to be genetically caused, and "secondary psychopaths," who are products of their environments.

The book that psychologists and psychiatrists use to categorize and diagnose mental illness, the DSM version IV, contains a category called "antisocial personality disorder" (APD).[1] This is a much broader category, which is behaviorally defined, and which includes both sociopathy and psychopathy. Roughly 1 in 5 people with APD is a psychopath (Kiehl and Buckholtz 2010). The percentage of psychopaths in the population as a whole is not known, but Kiehl estimates that between 15 and 35 percent of prisoners in the US are psychopaths (Kiehl and Buckholtz 2010).

The current approach to defining psychopathy and its related concepts is to use a list of criteria. The first such list was developed by Hervey Cleckley, who is known as the first person to investigate psychopaths using modern research techniques (Cleckley and Cleckley 1982). Anyone fitting enough of these criteria counts as a psychopath. There are several such lists in use. The most commonly used is called the Psychopathy Checklist Revised (PCL-R), which contains 20 questions (Hare 1991). An alternative version was developed in 1996 by Lilienfeld and Andrews, called the Psychopathic Personality Inventory (PPI) (Lilienfeld and Andrews 1996). In addition, the World Health Organization delineates a similar category it calls "dissocial personality disorder." The questions on these tests and their associated criteria pertain to both the behavior and personality of the interviewee. One future issue will be whether we want to shift the criteria for psychopathy from the more behaviorally oriented ones currently used to a brain-based definition: a psychopath is someone with this sort of brain. This assumes that the neuroscience of psychopathy establishes that a single type of brain is responsible for psychopathic behaviors, and that there are not several different variants.

While each of the inventories is different, there are significant areas of overlap, including the following criteria:

- *Uncaring*

  The PCL describes psychopaths as being callous and showing a lack of empathy, a trait which the PPI describes as "cold-heartedness." The criteria for dissocial personality disorder include a "callous unconcern for the feelings of others."

- *Shallow emotions*

  Psychopaths, and to a degree sociopaths, show a lack of emotion, especially the social emotions, such as shame, guilt, and embarrassment. Cleckley said that the psychopaths he came into contact with showed a "general poverty in major affective reactions," and a "lack of remorse or shame." The PCL describes psychopaths as "emotionally shallow" and showing a lack of guilt.

  There are now several lines of evidence that point to the biological grounding for the uncaring nature of the psychopath. For us, caring is a largely emotion-driven enterprise. The brains of psychopaths have been found to have weak connections among the components of the brain's emotional systems. These disconnects are responsible for the psychopath's inability to feel emotions deeply. They are also not good at detecting emotions in voices of other people, especially fear (Blair et al. 2002). They score poorly on tasks involving identifying emotions in faces. In consonance with their public image, psychopaths have reduced fear reactions: they show smaller reactions to the threat of impending electric shock. The emotion of disgust also plays an important role for our ethical sense. But psychopaths have extremely high thresholds for disgust. They show smaller reactions to the gruesome sight of mutilated faces, and to foul odors.

- *Irresponsibility*

  According to Cleckley, psychopaths are unreliable, while the PCL mentions "irresponsibility" and the PPI describes psychopaths as showing "blame externalization." The criteria for dissocial personality disorder include a disregard for social norms, rules, and obligations.

- *Insincere speech*

  Ranging from what the PCL describes as "glibness" and "superficial charm" to Cleckley's "untruthfulness" and "insincerity," all the way to "pathological lying," psychopaths devalue speech by inflating and distorting it toward selfish ends. The criteria for APD include "conning others for personal profit or pleasure." This casual use of words may be attributable to what some researchers call a "shallow sense of word meaning." Psychopaths do not show a differential brain response to emotional terms that normal people do (Williamson et al. 1991). They also have trouble understanding metaphors and abstract words.

- *Overconfidence*
  The PCL describes sociopaths as possessing a "grandiose sense of self worth."

- *Selfishness*
  Cleckley spoke of his psychopaths showing a "pathologic egocentricity [and incapacity for love]," which is affirmed in the PPI by its inclusion of egocentricity among its criteria. The PCL also mentions a "parasitic lifestyle."

- *Inability to plan for the future*
  Cleckley said that his psychopaths showed a "failure to follow any life plan." According to the PCL, psychopaths have a "lack of realistic long-term goals," while the PPI describes them as showing a "carefree nonplanness."

- *Violence*
  The criteria for dissocial personality include, a "very low tolerance to frustration and a low threshold for discharge of aggression, including violence." The criteria for antisocial personality disorder include "irritability and aggressiveness, as indicated by repeated physical fights or assaults." The PCL also describes psychopaths as having a need for stimulation and a proneness to boredom, which may be causal factors behind the violence.

8.3   Neuropsychological theories of psychopathy

8.3.1   The attentional model

According to Newman and his colleagues, the core deficit in psychopathy is a failure of what he calls "response modulation" (Hiatt and Newman 2006). When normal people engage in a task, we are able to alter our activity, or modulate our responses, depending on peripheral information. Psychopaths are specifically deficient in this ability, and according to Newman, this explains the impulsivity of psychopaths, as well as their problems with passive avoidance and with processing emotions.

Top-down attention tends to be under voluntary control, whereas bottom-up attention happens involuntarily. This is captured in the folk-psychological language of attention reports. In the case of top-down attention, we say "I looked for my car," whereas cases of bottom-up attention are expressed passively: "The sound drew my attention." Bottom-up attention can temporarily capture top-down attention, as when movement in the periphery of our visual field attracts our attention. Psychopaths have trouble using top-down attention to accommodate information that activates bottom-up attention during a task. In normal people, this process tends to happen automatically. When the hunter is scanning for deer, a rabbit hopping into the periphery of his visual field automatically attracts his attention.

The Stroop task is a neuropsychological test in which the subject must quickly state which color of ink a word is printed in. What makes the task difficult is that the words are themselves color words, so that for instance, the word "red" appears written in blue ink. Normal people have a slight delay as they have to inhibit the tendency to say "red" and instead answer "blue." There are now several studies indicating that psychopaths actually perform better than normal people on these tasks (Newman et al. 1997; Hiatt et al. 2004), presumably because they are not distracted in the way that normal people are. Once they have begun an activity, psychopaths are also insensitive to shifts in the pattern of rewards for their actions. In one study, psychopaths were shown a series of playing cards on a screen. They were told that they would receive one point for each face card but lose a point for each non-face card. The deck was deliberately stacked so that the first ten cards were face cards, then nine of the next ten, then eight of the next ten and so on. Subjects were told they could stop playing at any time. Non-psychopaths noticed the worsening trend and tended to stop playing after about 50 cards. Psychopaths played doggedly on, however, until the deck was almost finished and their winnings were gone.

8.3.2 The amygdala model

Blair and his colleagues have argued that the amygdala and ventromedial prefrontal cortex (vmPFC) are the core dysfunctional areas in psychopathy (Blair et al. 2005). The notorious lack of fear shown by psychopaths points to an amygdala dysfunction. In an fMRI study of fearful expression processing, Marsh and Blair reported reduced functional connectivity between the amygdalae and the vmPFC in children with psychopathic tendencies (Marsh and Blair 2008). Moreover, Birbaumer et al. (2005) reported reduced vmPFC activity as well as reduced amygdala activity in individuals with psychopathy during aversive conditioning (Birbaumer et al. 2005). Blair, Mitchell and Blair (2005) have argued that amygdala function is impaired in psychopaths, leading to dysfunctional creation and processing of affect-laden representations, particularly of others the psychopath may harm (Blair et al. 2005). In this regard, psychopaths may be similar to patients with damage to the ventromedial prefrontal cortex who are said to be suffering from "acquired sociopathy" (Roskies 2003). In persons with normal cognition, the vmPFC tends to take emotional input from the amygdala, and plays a role in anticipating and modulating rewards and punishments (Kringelbach 2005). Motzkin et al. (2011) found reduced functional connectivity between the vmPFC and amygdala in a sample of psychopaths. They also found reduced structural integrity of the right uncinate fasciculus, the primary white-matter connection between the vmPFC and the anterior temporal lobe including the amygdala, which they suggest is the ground of the reduced functional connectivity (Motzkin et al. 2011).

Roskies claims that vmPFC patients have normal reasoning capacities, but are simply not motivated to act on moral beliefs (Roskies 2006). This may be due to their inability to experience moral emotions such as empathy. When asked to provide an answer to the famous "trolley" thought experiment – where subjects are asked to decide whether to intentionally kill one person to save five – patients with ventromedial damage are more likely to judge that intentionally killing the one person (by pushing him onto trolley tracks) is the right thing to do, despite their having an active role in the killing. Thus, it is thought that persons with ventromedial damage may be more likely to engage in antisocial or immoral behavior, precisely because they do not feel badly about such actions.

When subjects are presented with moral dilemmas having a strong emotional character, activity in a network including the amygdala, medial prefrontal cortex, posterior cingulate gyrus, and the angular gyrus is observed. This network has been found to be active during "self-referential thinking, emotional perspective taking, recalling emotional experiences to guide behavior and integrating emotion in social cognition" (Glenn et al. 2009, 6). Glenn et al. found reduced activity in the amygdalae of psychopaths during emotional decision making, and found that a subgroup of these subjects who were skilled at conning and manipulation showed reduced activity within this "moral circuit." "Dysfunction in these regions," say Glenn et al., "suggest failure to consider how one's actions affect others, failure to consider the emotional perspective of the harmed other, or a failure to integrate emotion into decision making processes" (*Ibid.*). Similarly, Blair and Cipolotti (2000) extensively tested a subject with acquired sociopathy as a result of damage to his orbitofrontal cortex and left amygdala. They found that his responses to negative emotions were diminished, and they argue that his primary problem was an inability to be sensitive to others' anger.

8.3.3 The paralimbic model

Kiehl accepts that the amygdala is dysfunctional in psychopaths, but also implicates a much wider area of dysfunction, called the paralimbic cortex. This collection of cortical areas, which includes the anterior cingulate, posterior cingulate, superior temporal, insular and hippocampal cortex, forms a ring of inner cortical zones around the thalamus. Evidence that the insula is dysfunctional in psychopathy favors Kiehl's broadening of the areas of suspicion beyond the amygdala and ventromedial cortex. The insula, hidden in the fold that separates the temporal lobe from the lateral cortex above it, has been found to activate when the subject detects violations of social norms. It also activates when subjects experience anger, fear, empathy, and disgust (Kiehl and Buckholtz 2010). The insula also plays a role in pain perception, so dysfunction there may explain experimental findings in which psychopaths were insensitive to the threat of impending pain, in this case electric shock. The right insula and right hippocampus were also found to have smaller volumes in those scoring high on the PCL-R (Cope et al. 2012). A subsequent study on adolescents who scored highly on a version of the PCL-R adapted for younger people showed decreased gray matter volume in several paralimbic areas, including the orbitofrontal cortex, bilateral temporal poles, and the posterior cingulate (Ermer et al. 2013).

Cognition without the proper mix of emotion (or, more neutrally, autonomic activity) – whether it is too much or too little emotion – may be aimless and subject to being sidetracked by poor reasoning. The role of emotion in cognition goes beyond that of merely inhibiting us from doing harmful, illegal, or counterproductive things. It guides our reasoning, and can provide us with a sense of how strong or weak an argument is. For example, Baird and Fugelsang discovered that adolescents take longer to make moral decisions, partly because they have yet to develop the "gut" feeling which requires one to stop considering a decision when an immoral or harmful result is realized (Baird and Fugelsang 2004). Without this feeling, a strong reason to do x and a weak reason to not do x can appear to be equal. This can cause a sort of neutralizing effect, in which a weak argument and an opposing strong argument are taken to be equal in force.

8.3.4   The executive model

Morgan and Lilienfeld (2000) conducted a meta-analysis of the existing research on executive function in people diagnosed as exhibiting antisocial behavior, a large category that includes those diagnosed with antisocial personality disorder, as well as those diagnosed as psychopathic. They found that the antisocial behavior group scored .62 standard deviations worse on tests of executive function, which yielded a medium to large effect size. This included a finding of response perseveration in a group diagnosed as psychopathic (Newman et al. 1987; Yang et al. 2011). Since then, several attempts have been made to delineate subtypes within the category of psychopaths, partly in order to discern whether certain groups might have more severe executive function deficits. Recent research has distinguished two categories of psychopath who apparently have very different executive profiles: successful psychopaths, with little or no criminal record, and unsuccessful psychopaths, currently incarcerated or with a substantial criminal record. Gao and Raine (2010) recently published a review of studies distinguishing the two populations within five different samples: a community recruited sample, individuals from temporary employment agencies, college students, psychopaths employed in business and industry, and psychopathic serial killers. Unsuccessful psychopaths showed reduced prefrontal and amygdala volumes as well as hippocampal abnormalities, resulting in reduced executive functioning, including impaired decision-making (Gao and Raine 2010). Unsuccessful psychopaths also exhibit impaired autonomic/somatic responses and fear-conditioning deficits which are thought to contribute to poor and risky decision-making (Gao and Raine 2010). In contrast, successful psychopaths do not show similar structural and functional impairments of the prefrontal cortex, amygdala and hippocampus (Gao and Raine 2010).

Ishikawa et al. (2001) found that successful psychopaths actually had greater autonomic responses than both unsuccessful psychopaths and normal controls (as measured by their heart rate reactivity) during a task designed to produce embarrassment: preparing and then delivering a two-minute speech detailing their personal faults and weaknesses. The Ishikawa et al. study also found that, compared with unsuccessful psychopaths who had at least one criminal conviction, successful psychopaths had enhanced executive functioning as measured by the Wisconsin Card Sorting Task (WCST) (Ishikawa et al. 2001). The WCST is used to assess the following frontal lobe functions: strategic planning, organized searching, shifting of cognitive sets, considered attention, and modulating responses (Ishikawa et al. 2001). Indeed, successful psychopaths showed significantly better performance on the WCST than non-psychopathic controls (Ishikawa et al. 2001). In contrast, unsuccessful psychopaths scored lower than the controls, even though the two psychopathic groups did not differ on full scale IQ compared with the controls (Ishikawa et al. 2001). Ishikawa and colleagues suggested that better executive function might play a protective role for successful psychopaths, decreasing their tendency to be caught up in the criminal justice system (Ishikawa et al. 2001). This executive profile may also make successful psychopaths more effective at manipulating people.

8.4 Psychopathy, responsibility, and punishment

There has been much written about the criminal culpability of psychopaths in the past ten years, with many scholars arguing that psychopaths are at least partially excused from criminal responsibility. However, in most cases criminal courts have continued to deem psychopaths fully responsible. By some estimates there are half a million psychopaths currently in US prisons (Kiehl and Buckholtz 2010). Juries have been known to assign psychopaths the most serious of criminal punishments: In 2010, an Illinois jury sentenced murderer James Dugan to death, despite the defense's offering of expert psychological and neuroscientific evidence that he was a psychopath (Hughs 2010).

One can see the tension psychopathy generates for the law in the wake of high profile murder cases, such as the Sandy Hook Elementary School shootings, where many newspaper articles and blogs asked questions such as "Was Adam Lanza Sick or Evil?"[2] On the one hand, many consider any person who could kill first graders obviously sick; on the other hand, the instinct to call an obviously dangerous person who could commit such terrible crimes "evil" and punish him harshly is extremely strong. Many psychopathic offenders generate the same response: their cruel actions must mean they are sick, which would mitigate responsibility and punishment, or they are viewed as evil, which means they are more deserving of punishment. This tension is buoyed by the use of antisocial and criminal behavior in diagnoses of psychopathy, which could result in a lot of overlap between the category of "people who do really bad things" and "psychopath." For psychopathy to be an excuse under the law, it is not enough to recognize that psychopaths' behavior indicates they are abnormally antisocial and dangerous, for many dangerous people should be held fully responsible for their acts. Instead, to qualify for an excuse under the law, psychopaths must suffer from cognitive impairments significant enough to distinguish their decision-making and action from those of the normal responsible agent.

The disagreement among scholars regarding the responsibility of psychopaths appears to be related to a dispute about the decision-making capacities necessary for culpability. In turn, this dispute about mental capacity further reflects differences regarding the ultimate justification for criminal law and punishment, because the reasons why we punish affect our views of who should be punished.

8.4.1   The functions and justification of punishment

Criminal sanctions, including incarceration, are designed to serve particular functions. These are often called the principles of punishment, and there are four that are referred to most often: retribution, deterrence, incapacitation, and rehabilitation. According to the principle of retribution, violators of the law should get their "just deserts": punishment should serve to provide harmful consequences in response to a harmful act. The principle of deterrence attempts to influence an offender's decision-making with the threat of punishment. Both the general population and the specific offender who is punished are thought to be deterred from criminal acts by punishment. The principle of incapacitation also aims to stop defendants from offending, but there is no attempt to influence decision-making; instead the offender's environment is manipulated to make reoffending impossible, typically via incarceration. Finally, rehabilitation is the idea that offenders can be reformed so that they won't reoffend.

These *functions* of punishment are generally thought to fall into two broad categories of *justification* for punishment. A justification for punishment provides good reasons why society is warranted in denying offenders' liberties based upon their performance of certain acts. Traditionally deterrence, incapacitation, and rehabilitation were seen as *utilitarian* functions of punishment best understood and justified using a consequentialist theory. All three of these functions of punishment can be explained in terms of their consequences and an offender's rational tendency to maximize utility; or lack thereof (e.g. their failure to take into consideration societal-level utility). *Deontological moralism*, on the other hand, is clearly reflected in the principle of retribution, which states that violators of the law should get their "just deserts" in the name of moral notions of justice. This means that punishment should serve to provide harmful consequences in a response to a harmful act. Offenders ought to act out of duty to the moral law, and when they do not, they deserve moral condemnation and punishment proportional to the moral harm caused by their action.

Virtue theory represents a less popular third way punishment can be justified, by emphasizing the criminal law's obligation to "promote human flourishing by instilling and cultivating the moral virtues, promoting sound practical reasoning and punishing those who display vice" (Yankah 2009). Virtue theory is most obviously tied to punishment's function of rehabilitation, which aims to reform offender's characters such that they won't recidivate. However, the other functions of punishment can also be seen through the lens of virtue theory: as attempts to influence choices and character in the case of deterrence, incapacitation, and rehabilitation, or as moral judgment which refers to, and should be respectful of, character in the case of retribution.

As one legal scholar has noted, one of the central problems in the criminal law is that it cannot be justified by a single theory (Brown 2002). Because of this, attempts to make utilitarianism, or deontological theory, the sole justification for criminal law have been unsuccessful (Brown 2002). Despite recent changes to the Model Penal Code which seem to reflect an emphasis on retribution as the primary function of punishment, the current US criminal justice system seems to embrace multiple functions of punishment, and thus seems to require multiple justifications for its structure. Indeed, one might argue that the best version of the criminal justice system may be informed by all three and attempts to balance the four functions of punishment so as to produce social order and moral justice, and to promote good moral character.

8.4.2 The mental capacities necessary for criminal culpability

Many legal scholars pose questions of criminal culpability in terms of legal rationality expressed in the language of folk psychology. For example, Stephen Morse argues that the law's conception of the person as a practical reasoner is inevitable given the nature of the legal system: the law is meant to give people reasons to act, or refrain from acting, and hence requires that people be capable of acting for reasons. According to Morse, "It is sufficient for responsibility that the agent has the general capacity for rationality, even if the capacity is not exercised on a particular occasion" (Morse 2000, 253). In turn, the lack of a general capacity for rationality explains those cases where the law excuses persons from responsibility. Morse defines this general capacity as an underlying ability to engage in certain behavior. If a person is capable of certain conduct, it is fair to hold her responsible for failing to engage in such conduct.

Morse fleshes out his account by including the following capacities as constitutive of rationality: (1) the ability to perceive the world accurately, form true and justifiable beliefs; and (2) the ability to reason "instrumentally, including weighing the facts appropriately and according to minimally coherent preference-ordering" (Morse 2000, 255). Weird or abnormal desires themselves don't make a person irrational unless she lacks the rational capacities to weigh and order her desires. Therefore a person with disorders of desire is excused only where a desire is so strong and overwhelming that he loses the capacity to be guided by reason. Overall, the law's standard for rationality is set fairly low, according to Morse, because our legal system "has a preference for maximizing liberty and autonomy" (Morse 2000, 255).

H. L. A. Hart argued that the capacities necessary for responsibility may be "diminished" or "impaired," as well as wholly lacking, "and persons may be said to be 'suffering from diminished responsibility' much as a wounded man may be said to be suffering from a diminished capacity to control the movements of his limbs" (Hart 1968, 228). The defense of diminished capacity recognizes that some defendants may have decreased legal rationality or capacity, and allows a criminal defendant to reduce the degree of the crime for which he may be convicted, even if the defendant's conduct satisfies all the elements of a higher offense (Morse 1984; Morse 2003). Courts may also use the doctrine of diminished capacity to decrease the level of punishment. This "partial responsibility" application of diminished capacity is justified by the principle of proportionality, whereby punishment is moderated to be proportional to both the harm caused and the type of offender. Those who suffer from diminished capacity are thought to be less responsible for their acts because they do not have the capacity to form intentions in the way that normal adults do.

Interestingly, the three justifications for criminal law each seem to emphasize slightly different cognitive capacities as necessary for culpability under the law. Consequentialism highlights the need for rational capacities as a means to grasp and reflect upon the consequences of action. Virtue theorists similarly claim the practice of practical reason is necessary to develop character and exercise virtuous traits. However, also important to the practice of virtuous traits is the requirements that an actor feel the right way about her actions, and the permanence of personality traits which then dictate action. Deontologists also require that a responsible actor grasp the moral reasons for or against action, where this understanding often includes possessing the appropriate emotional responses to ethical situations.

For example, Oliver Wendell Holmes is considered a proponent of the consequentialist model. Holmes famously stated that the law ought to be understood and persuasive to the "bad man," who cared about nothing but his own interests (Holmes 1997). It seems possible that under Holmes' theory the bad man need not feel the right way about his acts, but only know they are forbidden, for him be responsible under the law. Morse, on the other hand, argues that legal rationality includes the capacity to act for moral reasons, claiming that "Unless an agent is able to understand what the victim will feel and is able to at least feel the anticipation of unpleasant guilt for unjustifiably harming another, the agent lacks the capacity to grasp and be guided by the primary rational reasons for complying with legal and moral norms" (Morse 2000). Thus Morse's theory would seem to be deontological, in that he believes a lack of emotional data regarding the potential consequences of one's act translates into a wholesale lack of legal rationality.

8.4.3 The legal capacity of psychopaths

Because of their lack of emotional data, Morse has argued that at least some psychopaths are not criminally responsible because they are thus not legally rational. Indeed, Morse argues for an extension of the current grounding conditions for legal insanity to include psychopathy (Morse 2008). Other philosophers have claimed psychopaths are not fully culpable because they lack personhood (Murphy 1972), or moral knowledge (Fields 1996). We have argued that some psychopaths' deficits in executive function mean they are not fully rational (Sifferd and Hirstein 2013). Again, these different positions on psychopathy as an excuse reflect differences of perspective on the constituents of legal rationality, which further reflect different justifying theories of criminal law.

On a traditional utilitarian theory of criminal law, the behavior of psychopaths can seem incomprehensible. Under utilitarianism, rationality was often portrayed as a "cold" process whereby an actor attempted to maximize utility. The threat of punishment, discounted by the likelihood the punishment will be imposed, was thus thought to dissuade at least some potential offenders from offending. As indicated above, this means utilitarian theories of law see social order, and thus deterrence, incapacitation, and rehabilitation, as the primary functions of punishment. The reason why some actors are not dissuaded from committing crimes may be that they discount the future (e.g. they discount the pain of punishment in light of the potential gain of the crime); they discount the likelihood of being caught; or they have nothing to lose from the crime (e.g. their current situation is as bad or worse than prison). The appropriate level of punishment is the amount necessary to outweigh the potential gains from the criminal act (Bentham 1789).

Although many psychopaths are of average or above average intelligence, they systematically fail to be persuaded by the threat of punishment such that they refrain from committing crimes. Thus if rationality is equated with intelligence, psychopaths are fully rational and yet commit crimes due to their narcissism or Machiavellianism (Paulhus and Williams 2002). According to the attentional model, some psychopaths may not be fully rational due to their deficits in attention, although it seems unlikely that attentional deficits alone, without other executive deficits, would be enough to place psychopaths beneath the very low bar of legal rationality. Similarly, dysfunction in fear or motivation are unlikely to result in a person's inability to maximize utility via something like means/end reasoning.

If one accepts the paralimbic model of psychopathy, then one might claim that psychopaths are missing correct emotional data, and then argue such data is crucial to rationality, using something like a Damasio-style somatic marker model of rationality (Damasio 1995). It may be that emotional feedback plays an important inhibitory role in ethical decisions, and without this feedback psychopaths cannot stop themselves from causing harm. From the utilitarian perspective, this would require an argument that psychopaths were so limited in their ability to go through the same rational process of weighing the costs and benefits of breaking the law that they have diminished mental capacity. It seems that the executive function model could provide a fuller account of the necessary tools for legal rationality – which are not just attentional, but involve access to memory, inhibition, use of theory of mind capacities, etc. – and also a diagnostic tool for determining what level of executive dysfunction may be excusatory.

But what about the "successful" psychopaths with intact executive functions? They have an emotional lack, but also the ability to reflect upon and inhibit their actions, despite their emotional lacks. From the utilitarian perspective, which sees persons as rational utility maximizers, and labels acts wrong when they result in harm and undermine social order, it seems that so-called successful psychopaths are fully culpable. Indeed, there is some evidence that at least certain groups of psychopaths are excellent utilitarian actors who aren't overly narcissistic: in one study psychopaths were significantly more likely to endorse harming others when commission of the harm would maximize aggregate welfare – the 'utilitarian' choice (Koenings et al. 2012). Even so, there is no question successful psychopaths are different from the average person, and that this difference requires special interventions and effort for the successful psychopath to be law-abiding. Even if successful psychopaths are rational in the eyes of the law, they are also actors who need special attention or assistance regarding their behavior. Just as a color blind motorist will need to take special rule-following precautions in order to obey traffic signals (such as memorizing the location, and not the color, of stop and go signals on traffic lights), so too do successful psychopaths need special help policing their behavior so it doesn't cause harm.

The psychopath's heightened tendency to cause harm may be more easily explained from the perspective of deontological moralism or even virtue theory, both of which emphasize the way in which an actor's moral feelings may guide their behavior as crucial to the moral quality of the act. All psychopaths fail to have normal emotional responses to cues of distress in that they lack some of the moral emotions which make salient potential harmful outcomes of behavior. While a utilitarian theory can attempt to include this information in their conception of rationality, a Kantian (deontological) moral actor, and a person with a good Aristotelian character, is required to feel the right way about an action for it to be the right action at all. Even if successful psychopaths are excellent utilitarian actors, capable of sophisticated reasoning with regard to the outcomes of their action, they still don't feel the right sorts of emotions about their actions, which results in a greater likelihood to underestimate the value of human life and interests when compared to normal actors.

Interestingly, a recent study found that participants who indicated greater endorsement of utilitarian solutions had higher scores on measures of psychopathy (Bartels and Pizarro 2011). The experimenters presented subjects with variants of the trolley problem – either watch five passengers in a runaway trolley car die, or push one bystander onto the tracks to his death to stop the car – and also asked questions to track their psychological dispositions, finding a strong link between the anti-social tendencies and willingness to kill the bystander to save the trolley passengers. The implication of the study was that appropriate moral feelings may lead one to take more seriously deontological commitments such as the categorical importance of human life or justice. Again, it seems that a deontological justification for punishment may have an easier time excusing the psychopath, given that from this perspective possessing the right sorts of moral emotions is so central to doing the right thing. In this sense, Morse's theory of psychopathy, which argues that the psychopath's lack of moral emotions such as empathy and guilt make him so irrational as to be legally insane, can be seen as compatible with a deontological theory of law and punishment.

8.4.4　The punishment of psychopaths

From a practical standpoint, if a psychopathic offender is deemed eligible for the excuse of diminished capacity, he may then qualify for a lesser crime or less severe punishment (just as a severely mentally retarded offender may be found guilty of manslaughter, instead of first degree murder, or deemed ineligible for the death penalty). If the psychopathic offender is given a shorter sentence, this result is worrying because the cognitive incapacity that qualifies the psychopath for an excuse of diminished capacity is likely to make him likely to recidivate. Indeed, this quite serious worry may be part of what motivates Morse to claim that psychopaths should be considered for the insanity plea: offenders deemed legally insane are incapacitated in a hospital for the mentally ill, often for longer than their criminal sentence would have been had they been convicted (Perlin 1994).

However, the future dangerousness of a defendant is not relevant at the guilt phase of a trial, which aims only to determine guilt regarding a particular crime. At sentencing, future dangerousness may in some cases be considered (e.g. in capital cases); but at the same time, diminished capacity may be considered as a mitigating factor, cancelling out the aggravating factor of dangerousness. In the end, however, the hard case of psychopaths does not seem to be a good reason to alter the traditional handling of the defense of diminished capacity. As legal scholars know well, hard cases make bad law. Altering the structure or application of the diminished capacity excuse to fit the psychopathic offender threatens the justice and coherence of the law. Instead, the criminal justice system might attempt to address concerns about recidivism by offering rehabilitative programming to psychopathic prisoners – intensive cognitive therapy has been shown to have some effect on antisocial behavior by psychopaths (Skeem et al. 2002) – or by subjecting psychopathic offenders to specific conditions for parole.

The latter possibility, of strict monitoring and reporting requirements for psychopaths upon release, is probably more realistic given the cost of intensive cognitive therapy. Despite worries about its ability to predict dangerousness, the PCL-R is already used in many US jurisdictions to inform parole decisions (Hare 1998). The psychopathic parolee could be subject to something like a registration program, similar to that many sexual offenders are required to undergo; although these sex offender registration programs make clear that there are significant risks in publicly tagging offenders as dangerous.

8.5 A sociopathic society?

According to David Lykken, one of the primary researchers of sociopathy and psychopathy, our society has become an incubator for sociopaths. They "occur in ever-increasing numbers, especially in our cities," he says, where we are producing sociopaths "with factory like efficiency and at enormous cost" (Lykken 1995, 7). These environmental influences work in conjunction with what appear to be genetic roots of psychopathy. In a study of seven-year-old twins, Viding et al. (2005) found that one of the core traits of psychopathy that manifests itself early in life, the tendency to be callous and unemotional, was under "strong genetic influence." The possibility that the human race contains members genetically programmed to sabotage our attempts to create an ethical society is disquieting. Given the unguided nature of evolution, it is plausible that a phenomenon like this could arise. Many male psychopaths are adept at seducing women, and this guarantees that they will pass on their genes. One way to prevent the percentage of psychopaths in a society from rising is to sensitize ourselves to their characteristics and their consequences. We fear that this is currently not happening, and that, at least in the US, we are creating a society that in many ways admires and nurtures psychopaths.

The psychologist Robert Hare is acknowledged as the foremost expert in the psychological characteristics of sociopaths. In recent writings with colleagues (Babiak and Hare 2009; Babiak et al. 2010), he has begun to suggest that certain corporation presidents might be considered psychopathic because of their behavior toward their consuming public. There is evidence that some businesspeople do not care about who their products hurt, as long as they are profitable. They are able to get away with this sort of behavior partly because of a social climate that is favorable toward them. A case that has become emblematic of this unconcern for human life happened in the 1970s at Ford Motor Corporation. After it became apparent that the gas tank design of the hugely popular Ford Pinto was dangerous, a calculation was made by the Ford Motor Corporation of how many people their defective gas tank would kill (after exploding due to an impact from the rear), and how much money they would lose on each lawsuit vs. how much it would cost to fix the gas tanks. The projected lawsuit losses were less than the cost of fixing the cars, so the decision was made not to fix them, and several more people burned to death as a result. The person at the head of the table during these discussions was Lee Iacocca, who later became a famous icon of business, and whose books dispensing management advice were widely read.

8.6   Conclusion

The emerging neuroscience of psychopathy will have several important implications. In this paper we reviewed four competing neuropsychological theories of psychopathic cognition. The first of these models, Newman's attentional model, locates the problem in a special type of attentional narrowing that psychopaths have shown in experiments. The second and third, Blair's amygdala model and Kiehl's paralimbic model, focus on the psychopath's emotional deficits, both in experiencing his own emotions as well as responding to the emotions of others. The fourth model locates the problem at a higher level, and may be better able to account for the heterogeneous nature of the psychopathy. This model accounts for the failure of psychopaths to notice and correct for their attentional or emotional problems using executive processes. Executive processes are a vital component of human rationality since they are responsible for planning actions, or inhibiting unwise actions, as well as allowing emotions to influence cognition in the proper way. Some successful psychopaths may have these abilities, while unsuccessful psychopaths may not.

We have evaluated psychopaths in light of the three primary theories used to justify criminal punishment: utilitarianism, deontological theory, and virtue ethics. Each emphasizes slightly different cognitive capacities as necessary for culpability under the law. Consequentialism highlights the need for rational capacities as a means to grasp and reflect upon the consequences of action. Virtue theorists similarly claim that practical reason is necessary to develop character and exercise virtuous traits. Deontologists require a responsible actor grasp the moral reasons for or against action, where this understanding often includes possessing the appropriate emotional responses to ethical situations.

The psychopath's heightened tendency to cause harm may be more easily explained from the perspective of deontological or virtue theory, precisely because they focus upon the way an actor's feelings are crucial to the moral quality of the act. All psychopaths fail to have normal emotional responses to cues of distress in that they lack some of the moral emotions which make salient potential harmful outcomes of behavior. Utilitarian theory can attempt to include this information into their conception of rationality; however, a deontological moral actor, and a person with a good character, is required to feel the right way about an action for it to be the right action at all. Thus it seems that deontological or virtue theory may have an easier time excusing the psychopath than utilitarian theory.

If a psychopathic offender is deemed eligible for a criminal excuse, he may then qualify for a lesser crime or less severe punishment. This is worrying due to the psychopath's heightened tendency to recidivate. The criminal justice system might attempt to address this problem by offering rehabilitative programs to psychopathic prisoners, or by subjecting psychopathic offenders to specific conditions for parole. We feel that it may be especially important for courts to strictly monitor psychopaths.

That old saw about whether humans are essentially good or essentially bad may be informed by our broadening knowledge of psychopaths, especially given the possibility that their condition may be at least partly genetic. Anyone defending the notion that we are basically good will need to add a rejoinder about how that generalization cannot include all of us. How the rest of us deal with the psychopathic population will determine the sort of future societies we will live in.

Notes

1. There is now a DSM version V, but it remains controversial to the point of being rejected by the US National Institute of Mental Health.
2. Kerwick (2012).

References

Babiak, P. and R. D. Hare (2009). *Snakes in Suits: When Psychopaths Go to Work.* New York: HarperCollins.

Babiak, P., C. S. Neumann, et al. (2010). Corporate Psychopathy: Talking the Walk. *Behavioral Sciences & the Law* 28(2), 174–193.

Baird, A. A. and J. A. Fugelsang (2004). The Emergence of Consequential Thought: Evidence from Neuroscience. *Philosophical Transactions of the Royal Society* 359(1451), 1797–1804.

Bartels, D. M and D. A. Pizzaro (2011). The Mismeasure of Morals: Antisocial Personality Traits Predict Utilitarian Responses to Moral Dilemmas. *Cognition* 121(1), 154–161.

Bentham, J. (1789). *An Introduction to the Principles of Morals and Legislation.* Oxford: Clarendon Press.

Birbaumer, N., R. Veit, et al. (2005). Deficient Fear Conditioning in Psychopathy: A Functional Magnetic Resonance Imaging Study. *Arch Gen Psychiatry* 62(7), 799–805.

Blair, R. J. R., D. Mitchell, et al. (2005). *The Psychopath:Emotion and the Brain.* Malden, MA: Jon Wiley & Sons.

Blair, R. J. R., D. G. V. Mitchell, R. A. Richell, S. Kelly, A. Leonard, C. Newman, and S. K. Scott (2002). Turning a Deaf Ear to Fear: Impaired Recognition of Vocal

Affect in Psychopathic Individuals. *Journal of Abnormal Psychology*, 111(4), 682–686.

Blair R. J. R. and L. Cipolotti (2000) Impaired Social Response Reversal: a Case of "Acquired Sociopathy." *Brain* 123(6), 1122–1141.

Brown, D. K. (2002). What Virtue Ethics Can Do for Criminal Justice: A Reply to Huigens. *Wake Forest L. Rev.* 37, 29.

Cleckley, H. M. (1982). *The Mask of Sanity*. New York: New American Library.

Cope, L. M., M. S. Shane, et al. (2012). Examining the Effect of Psychopathic Traits on Gray Matter Volume in a Community Substance Abuse Sample. *Psychiatry Research: Neuroimaging* 204(2), 91–100.

Damasio, A. (1995). *Descartes' Error: Emotion, Reason, and the Human Brain*. New York: Harper Perennial.

Ermer, E., L. M. Cope, et al. (2013). Aberrant Paralimbic Gray Matter in Criminal Psychopathy. *Journal of Abnormal Psychology* 121(3), 649.

Fields, L. (1996). Psychopathy, Other-Regarding Moral Beliefs, and Responsibility. *Philosophy, Psychiatry, & Psychology* 3(4), 261–277.

Gao, Y. and A. Raine (2010). Successful and Unsuccessful Psychopaths: a Neurobiological Model. *Behavioral Sciences & the Law* 28(2), 194–210.

Glenn, A. L., A. Raine, et al. (2009). The Neural Correlates of Moral Decision-Making in Psychopathy. *Mol Psychiatry* 14(1), 5–6.

Hare, R. D. (1991). *Manual for the Revised Psychopathy Checklist* (1st edn) Toronto, Ontario, Canada: Multi-Health System.

Hare, R. D. (1998). The Hare PCL-R: Some Issues Concerning Its Use and Misuse. *Legal and Criminological Psychology* 3(1), 99–119.

Hart, H. (1968). *Punishment and Responsibility: Essays in the Philosophy of Law*. Oxford: Clarendon Press.

Hiatt, K. D. and J. P. Newman (2006). Understanding Psychopathy: The Cognitive Side. In C. J. Patrick (Ed.) *Handbook of Psychopathy* (pp. 334–352). New York, NY: Guilford Press.

Hiatt, K. D., W. A. Schmitt, et al. (2004). Stroop Tasks Reveal Abnormal Selective Attention Among Psychopathic Offenders. *Neuropsychology* 18(1), 50.

Holmes, O. W. (1997). The Path of the Law. *Harvard Law Review* 110(5): 991–1009.

Hughes, V. (2010). Head Case. *Nature* 464, 340–342.

Ishikawa, S. S., A. Raine, et al. (2001). Autonomic Stress Reactivity and Executive Functions in Successful and Unsuccessful Criminal Psychopaths from the Community. *Journal of Abnormal Psychology* 110(3), 423–432.

Kerwick, J. (2012). Was Adam Lanza Sick Or Evil? *American Daily Herald*. December 19. http://americandailyherald.com/pundits/jack-kerwick/item/was-adam-lanza-sick-or-evil

Kiehl, K. and J. Buckholtz (2010). Inside the Mind of a Psychopath. *Scientific American*. September/October, 22–28.

Koenings, M., M. Kruepke, J. Zeier, and J. P. Newman (2012) Utilitarian Moral Judgement in Psychopathy. *Social Cognitive and Affective Neuroscience* 7(6), 708–714.

Kringelbach, M. L. (2005). The Human Orbitofrontal Cortex: Linking Reward to Hedonic Experience. *Nat Rev Neurosci* 6(9), 691–702.

Lilienfeld, S. O. and B. P. Andrews (1996). Development and Preliminary Validation of a Self-Report Measure of Psychopathic Personality Traits in Noncriminal Population. *Journal of Personality Assessment* 66(3), 488–524.

Lykken, D. T. (1995). *The Antisocial Personalities*, Hillsdale, NJ: Erlbaum.

Marsh, A. A. and R. J. R. Blair (2008). Deficits in Facial Affect Recognition Among Antisocial Populations: a Meta-Analysis. *Neuroscience & Biobehavioral Reviews* 32(3), 454–465.

Morgan, A. B. and S. Lilienfeld (2000). A Meta-Analytic Review of the Relation between Antisocial Behavior and Neuropsychological Measures of Executive Function. *Clinical Psychology Review* 20(1), 113–136.

Morse, S. J. (1984). Undiminished Confusion in Diminished Capacity. *Journal of Criminal Law and Criminology* 75(1), 1–55.

Morse, S. J. (2000). Rationality and Responsibility. *Southern California Law Review* 74, 251–268.

Morse, S. J. (2003). Inevitable Mens Rea. *Harvard Journal of Law & Public Policy* 27, 51–64.

Morse, S. J. (2008). Psychopathy and Criminal Responsibility. *Neuroethics* 1(3), 205–212.

Motzkin, J. C., J. P. Newman, et al. (2011). Reduced Prefrontal Connectivity in Psychopathy. *Journal of Neuroscience* 31(48), 17348–17357.

Murphy, J. G. (1972). Moral Death: A Kantian Essay on Psychopathy. *Ethics* 82: 284.

Newman, J. P., C. M. Patterson, et al. (1987). Response Perveration in Psychopaths. *Journal of Abnormal Psychology* 96, 145–148.

Newman, J. P., W. A. Schmitt, et al. (1997). The Impact of Motivationally Neutral Cues on Psychopathic Individuals: Assessing the Generality of the Response Modulation Hypothesis. *Journal of Abnormal Psychology* 106(4), 563.

Partridge, G. E. (1930). Current Conceptions of Psychopathic Personality. *American Journal of Psychiatry* 10, 53–99.

Paulhus, D. L. and K. M. Williams (2002). The Dark Triad of personality: Narcissism, Machiavellianism, and psychopathy. *Journal of Research in Personality* 36(6), 556–563.

Perlin, M. L. (1994). *The Jurisprudence of the Insanity Defense*. Durham, NC: Carolina Academic Press.

Pritchard, J. C. (1835). *A Treatise on Insanity*. London: Sherwood, Gilbert, and Piper.

Roskies, A. L. (2003). Are Ethical Judgements Intrinsically Motivational? Lessons from "Acquired Sociopathy." *Philosophical Psychology* 16(1), 51–66.

Roskies, A. L. (2006). Patients With Ventromedial Frontal Damage Have Moral Beliefs. *Philosophical Psychology* 19(5), 617–627.

Rush, B. (1812). *Medical Inquiries and Observations Upon the Diseases of the Mind*. Philadelphia: Kimber and Richardson.

Sifferd, K. L. and W. Hirstein (2013). On the Criminal Culpability of Successful and Unsuccessful Psychopaths. *Neuroethics*, 1–12.

Skeem, J., J. Monahan, et al. (2002). Psychopathy, Treatment Involvement, and Subsequent Violence Among Civil Psychiatric Patients. *Law and Human Behavior* 26(6), 577–603.

Williamson, S., T. J. Harpur, et al. (1991). Abnormal Processing of Affective Words by Psychopaths. *Psychophysiology* 28(3), 260–273.

Viding, E., R. J. R. Blair, T. E. Moffitt, and R. Plomin (2005). Evidence for Substantial Genetic Risk for Psychopathy in 7-Year-Olds. *Journal of Child Psychology and Psychiatry* 46(6), 592–597.

Yang, Y., A. Raine, et al. (2011). Abnormal Structural Correlates of Response Perseveration in Individuals with Psychopathy. *Journal of Neuropsychiatry and Clinical Neuroscience* 23(1), 107–110.

Yankah, E. N. (2009). Virtue's Domain. *U. Ill. L. Rev.*, 1167–1212.

9
Memory Traces, Memory Errors, and the Possibility of Neural Lie Detection

Sarah K. Robins

9.1 Introduction

The civic response to crime is forever challenged by the human inclination to lie, deceive, mislead, and distort. There is thus perennial interest in creating lie detectors, which can distinguish between honest and deceptive reports during interrogation. Most lie detectors work by identifying behavioral or physiological correlates of deception. The traditional polygraph test, for example, measures a person's heart rate, skin conductance, and blood pressure (amongst other physiological markers) while he or she answers questions, detecting the elevated arousal that often accompanies deception. The problems with such measures are well known: the connection between arousal and deception is imperfect, and respondents can develop tactics by which the connection is further weakened or suppressed. Many are excited by advances in neuroscience, which suggest the possibility of neural lie detection (e.g., Spence et al. 2001), as these are thought to provide more direct and reliable measures of guilt and deception.

Recently, the company *Government Works, Inc.* has begun promoting a technology called Brain Fingerprinting, which is advertised as a way to detect guilt in the brain. According to its inventor, Lawrence Farwell, Brain Fingerprinting works not by detecting lies, but by detecting memories (Farwell & Donchin 1991). This technology is an application of the neuroscientific discovery that people's electroencephalography (EEG) records show a characteristic positive response – the P300 – to familiar stimuli, but not to novel ones (Polich & Kok 1995). During Brain Fingerprinting, suspects are presented with information about a crime that only its perpetrator could know. If the presentation generates an elevated P300, this is offered as evidence of guilt. If not, the suppressed response is offered as evidence of innocence. Brain Fingerprinting has been used to convict (Farwell 2012) and exonerate (Harrington v. State 2001) suspects.

Others can determine whether it is ethical to subject suspects to Brain Fingerprinting and whether this test meets the evidentiary standards of legal admissibility. My focus is on the theoretical assumptions behind this technology. Specifically, I evaluate an assumption that has gone unquestioned by both proponents and critics of Brain Fingerprinting – the assumption that the P300 is a measure of recognition. Characterizing Brain Fingerprinting in this way – as a form of memory detection – requires commitment to an Archival View of Memory, according to which the brain stores memory traces of particular past events. Even critics of this technology, such as Meegan (2008), allow this view of memory to influence the kinds of errors they consider in their critique. The result is collective oversight of a particular form of memory error: false recognition. This is a mistake. False recognition is not only possible, it is a prevalent and persistent form of memory error. What's worse, instances of false recognition show a P300 response indistinguishable from instances of genuine recognition. False recognition creates problems for Brain Fingerprinting and the view of memory on which it is based. These limitations must be fully understood before the use of Brain Fingerprinting in law enforcement is given any further consideration.

I begin in Section 2 with an overview of Brain Fingerprinting's measures and methods. In section 3, I articulate the Archival View of Memory on which this technology is based, showing how the Archival View influences claims about the nature of recognition and error in Brain Fingerprinting. Section 4 is devoted to evidence of false recognition results, and Section 5 highlights two problems these results raise for Brain Fingerprinting.

9.2   The Brain Fingerprinting method

Brain Fingerprinting couples a standard way of measuring neural activity in cognitive neuroscience, the EEG, with a common interrogation tactic used in law enforcement, the Guilty Knowledge Test. I explain each of these in more detail below, and then describe how they are combined in Brain Fingerprinting tests.

9.2.1   EEG, ERP, and the P300

Electroencephalography (EEG) is a method of functional brain scanning that works by measuring electric signals from the brain via electrodes placed along the scalp.

Compared to other methods for detecting brain function – such as fMRI and PET – EEG offers poor spatial resolution. The electrodes placed on the scalp aggregate the electric signal from neurons across broad sections of the brain. The advantage of EEG is its temporal resolution: these electrodes can detect millisecond-by-millisecond changes in neural activity (Niedermeyer & da Silva 2004). EEG recordings are thus used to measure various rhythmic patterns in neural activity.

EEG can also be used to measure changes in neural activity that occur in response to a stimulus. These are known as Evoked Response Potentials, or ERPs. The ERP waveform has several distinct components, identified by their 1) polarity – whether the change in electrical signal is positive or negative – and, 2) onset – when, after the stimulus, the change occurs. Neuroscientists gain insight into brain function by observing how the components of the standard ERP waveform are altered by the presentation of different types of stimuli.

Brain Fingerprinting focuses on the P300 component of the ERP: a positive change in the ERP waveform that occurs approximately 300 milliseconds after the stimulus is presented. Farwell and colleagues refer to this component as the P300 *MERMER*: The Memory and Encoding Related Multifaceted Electroencephalographic Response.[1] The P300 is considered a cognitive component of the ERP waveform, as it is sensitive to differences in stimuli across various perceptual modalities (i.e., visual, auditory, olfactory; see Polich & Kok 1995 for review).

The P300 is the ERP component of focus in discrimination tasks, such as the Oddball Paradigm, where participants are instructed to keep track of two types of stimuli but to respond only to one of them, which occurs relatively infrequently (e.g., Verleger & Berg 1991). In this paradigm, presentation of the less frequent, oddball stimulus results in an amplified P300 response. The P300 has also been the focus of memory tasks, where it has been shown to be a reliable way of differentiating between new and old stimuli. Specifically, studies have shown that the P300 response is larger (i.e., the positive peak of the waveform at 300ms is higher) when the stimulus is one that the participant has seen before, compared to a stimulus that is novel for the participant (Senkfor & Van Petten 1998; Friedman & Johnson 2000).

9.2.2   Guilty Knowledge Test

The Guilty Knowledge Test (GKT) is an interrogation method used by law enforcement officials to identify information that a suspect knows about a crime, but may be trying to conceal (Lykken 1959). As with forms of interrogation based on lie detection, the GKT looks for physiological markers of arousal that may uniquely identify the crime's perpetrator(s). But unlike tests that try to establish whether a suspect is lying or being deceptive, the GKT evaluates how a suspect responds to information about the crime that law enforcement officials already know, but have not released to the public.

A polygraph test might, for example, include questions about the suspect's alibi, whereas a GKT would ask the suspect to identify the weapon used in the crime. A standard GKT is multiple choice, with the correct answer hidden amongst a set of plausible alternatives. If the suspect is innocent, then there should be no difference in the suspect's physiological response to each item presented. If the suspect shows an elevated physiological response to the correct answer (e.g., accelerated heart rate or increased pupil dilation), then this response is taken as evidence of guilt (e.g., Lubow & Fein 1996). The GKT has not been used as widely as the polygraph, but the approach has been gaining in popularity recently amongst law enforcement officials working in counterterrorism (Ben-Shakhar 2012).[2]

9.2.3   Brain Fingerprinting

Brain Fingerprinting is a version of the GKT, which uses EEG records to determine what a suspect knows about the crime by measuring differences in the ERP waveform – specifically, the amplitude of the P300 – when the suspect is presented with each possible answer to the multiple choice question. Possible answers come in three types: targets, irrelevants, and probes (Farwell & Donchin 1991; Farwell & Smith 2001). Targets are pieces of information about the crime that investigators know are known to the suspect, for example, details about the crime that were widely circulated in the media.[3] Irrelevants are the opposite: they are pieces of information that investigators know that the suspect does not know. In most cases, these are "details" about the crime that the investigators have made up. Probes are the stimuli of interest. They are pieces of information about the crime that a suspect could know only if he or she was present when the crime was committed. The targets, irrelevants, and probes may be presented to the suspect as words or phrases, or in some cases they may be pictures of the crime scene or objects used in the crime.

Brain Fingerprinting relies on the well-established difference in the P300 component of the ERP between stimuli that are familiar and stimuli that are novel. Recall that familiar stimuli produce a P300 response that is more amplified than the response to novel items. In this experimental setup, targets function as the familiar stimuli; the irrelevants are novel. By measuring the amplitude of the P300 waveform for targets and irrelevants, investigators can establish baseline measures for how a suspect responds to old and new pieces of information, respectively. They can then ask the crucial question: does the suspect's response to probes look more similar to that of targets or irrelevants? If the suspect's EEG records show an amplified P300 response to probes – information about the crime that only a perpetrator or witness could know – then this is taken as evidence of his or her involvement in the crime. If no such amplified response is detected, this is taken as evidence that the suspect was not involved. This method of interrogation is touted as highly reliable; laboratory tests of Brain Fingerprinting report error rates of less than 1% (e.g., Farwell & Richardson 2006; Farwell, Richardson, & Richardson 2012).

This technology has also been applied outside of the laboratory, to some real-world criminal investigations. Evidence from Brain Fingerprinting aided in the murder conviction of J. B. Grinder. Farwell's demonstration that Grinder's P300 responses to probes matched his P300 responses to targets led to Grinder's confession and subsequent sentencing (Farwell 2012). Brain Fingerprinting has also been used as evidence of a suspect's innocence. Terry Harrington spent more than two decades in prison, serving a life sentence for a murder he insisted he had not committed. Evidence from his Brain Fingerprinting test, which showed that his responses to probes matched his responses to irrelevants, was ruled legally admissible and played a role in his eventual release (Harrington v. State 2001).

9.3   Brain Fingerprinting as memory detection

Having sketched the Brain Fingerprinting method in the previous section, I turn now to an evaluation of its underlying theoretical assumptions. Farwell's description of this technology as a form of fingerprinting is intentional. Much as fingerprints are used to match a suspect to a particular location, Brain Fingerprinting is described as a measure that "matches information stored in the brain of the suspect with information from the crime scene" (Farwell et al. 2012, 118). This characterization of Brain Fingerprinting – as a form of memory detection – relies on a particular view of memory. Indeed, claims about memory play a central, if unarticulated, role in this approach to interrogation. It is thus important to understand what these commitments are and whether they can be supported.

In what follows, I argue that Farwell and other proponents of Brain Fingerprinting are committed to an Archival View of Memory, according to which the brain stores memory traces of particular past events. This assumption about the structure of memory is critical for their characterization of the P300 response as a measure of recognition and, in turn, the claim that this response can be used to identify information that is stored in the minds of criminals. Even those who have been critical of Brain Fingerprinting, such as Meegan (2008), accept the Archival View of Memory upon which this method of interrogation is based. As a result, both supporters and critics of Brain Fingerprinting have overlooked or downplayed the possibility of false recognition – an amplified P300 response to a stimulus in the absence of corresponding information in the brain.

9.3.1   Archival View of Memory

The Archival View of Memory holds that memory is a capacity by which we form, store, and reactivate detailed mental representations – memory traces – of particular past events. The Archival View has been the default view of memory throughout much of the history of philosophy and psychology. It is often traced to Aristotle and his characterization of memory as the preservation of perception, akin to the impressions a signet makes into a wax tablet (see Sorabji 2003). The view persists in Locke's characterization of memory traces as ideas in the mind's storehouse and Hume's claim that remembering is the revival of previous impressions.[4] Others have likened memory traces to phonographic records, images produced by a camera obscura, entries in a dictionary, and files in a cabinet.[5] These metaphors indicate how individual traces carry information as well as how these traces are organized in the memory store. If traces are pictures, then memory is the photo album in which they are displayed. If traces are words inscribed on a page, then memory is the notebook that contains each entry.

Farwell is committed to this Archival View of Memory. Brain Fingerprinting is promoted as a way to "unlock the truth within an individual's memory" (www.brainwavescience.com). Of course, this technology is focused on memories of only a small subset of past events – events where crimes were committed. Nonetheless, it is assumed that detailed traces of these events are contained in memory: "This technology [Brain Fingerprinting] evaluates the presence or absence of information/evidence in the one place where a comprehensive record of every crime is stored – the brain of the perpetrator" (Farwell & Smith 2001, 135).

How one thinks about the structure of memory has implications for how one thinks about 1) the act of remembering and 2) the kinds of memory errors that are possible. In the case of Brain Fingerprinting, this results in the following two claims:

1) The P300 response is an indication of recognition, and
2) False recognition is, if not impossible, a highly improbable memory error.

I consider each of these claims in more detail below.

9.3.2   Remembering, retrieval, and recognition

For proponents of the Archival View, the act of remembering is most often an act of retrieval. Much as one might fetch a book from the library by jotting down its call number and walking through the stacks to reach the desired location, so too remembering is characterized as the act of scanning one's mental repository to locate the desired piece of information. Homage to this claim can be seen in historical accounts of memory, as well as in contemporary computational models of memory from cognitive science. William James, for example, likens memory to a house with many rooms, characterizing remembering as a process by which "we search in our memory for a forgotten idea just as we rummage our house for a lost object" (1890: 654). Similarly, Anderson's ACT-R model of cognitive processing operationalizes remembering as a scan through a mental database, following a set of if-then search rules (Anderson 1976; Anderson, Bothell, Lebiere, & Matessa 1998).

The Archival View allows for remembering without retrieval in cases of recognition. Retrieval requires a deliberate, effortful search of the memory store, initiated in response to a cue or prompt. Recognition, in contrast, is automatic and effortless; it is a response that occurs when a person encounters something that matches an item stored in memory. Proust's (1913) description of tasting a madeleine that floods him with memories of his childhood is a clear example. In this way, recognition provides direct access to the contents of memory. For proponents of the Archival View, "recognition tests are believed to bypass the retrieval stage and to provide pure measures of what is stored in memory" (Roediger 2000: 56).

Farwell's confidence in Brain Fingerprinting is driven by his interpretation of the P300 as a measure of recognition:

> When an individual recognizes something as significant in the current context, he experiences an "Aha!" response. This response is characterized by a specific brainwave pattern known as a P300-MERMER. (Farwell 2012, 115)

Further, we can see that Farwell considers the P300 a pure measure of memory. He often describes Brain Fingerprinting as an "objective" and "scientific" form of interrogation, one that is immune to the problems that have plagued other forms of interrogation. On the Brain Fingerprinting website he explains its superiority to the polygraph and other versions of the GKT as a reflection of the fact that this test "detects *only information*" (www.brainwavescience.com /how_it_works.asp, emphasis in original).

9.3.3 Memory errors

Farwell and colleagues consider Brain Fingerprinting to be nearly immune to error. They characterize the P300 response as "100% accurate" (Farwell & Smith 2001, 141) and when urged toward modesty merely repackage this as a claim that Brain Fingerprinting yields "less than 1% error rates" (Farwell et al. 2012, 116).

Endorsing the Archival View does not compel the claim that memory is infallible. It does, however, influence the kinds of memory errors that one considers to be possible. The Archival View makes room for two general kinds of error: forgetting and retrieval failure. Forgetting occurs when a memory trace for a particular past event is missing from the memory store, either because it was lost/removed or never created in the first place. Retrieval failures are errors in how the memory store is searched. In these cases, the trace exists but is not found.

This view of memory errors shapes evaluations of Brain Fingerprinting's reliability, which – given the confidence expressed by Farwell and colleagues – is largely conducted by critics of this method of interrogation. These evaluations are often framed in terms of false positives and false negatives (e.g., Meegan 2008). A false positive would be a case where an innocent suspect produces a response consistent with guilt (i.e., generating a P300 for probes that matches that of targets, even though the stimulus is novel), whereas a false negative would be a case where a guilty suspect produces a response consistent with innocence (i.e., generating a P300 for probes that matches that of irrelevants, even though the suspect has previously encountered the stimulus). Since the P300 is a measure of recognition, circumventing the process of retrieval, there is no need to consider the possibility of memory errors generated by retrieval failure. The only remaining form of memory error is forgetting. If a guilty suspect forgot details of the crime, then this could result in a false negative, as the suspect's P300 responses would treat the stimuli as novel.

Meegan (2008) offers a set of reasons to be concerned about Brain Fingerprinting's susceptibility to false negatives. Crimes often occur in situations that could be inhospitable to forming detailed traces, which could mean that the suspect encodes no record of the event. This would result in a lack of stored information for the ERP to detect, despite the suspect's previous encounter with the information. Further, there is typically a long interval between when the crime is committed and when the suspect is interrogated. This, Meegan claims, opens up the possibility that a perpetrator's mental record of the crime will deteriorate, and thus fail to produce the amplified P300 response indicative of guilt. Given the likelihood of false negatives, Meegan is especially critical of the use of Brain Fingerprinting to support exoneration. Other evaluators of Brain Fingerprinting have found this line of argument compelling, even though they are more optimistic about the overall merits of this technology (e.g., Fox 2008; Iacono 2008).

Meegan (2008) is far less concerned about the possibility of false positives. He shares in Farwell's assumption that recognition is a pure measure of memory. This reflects a shared commitment to the Archival View, according to which recognition is an automatic, effortless response produced by a direct match between the stimulus and the memory trace. On this view, it is difficult to imagine how a recognition response could be produced in response to a stimulus if there were no such stored information. Meegan thus characterizes false recognition as, at best, infrequent – a situation that could occur only in highly contrived experimental conditions. As he states:

> False feelings of familiarity are a relatively rare occurrence in everyday life – it is not as if objects we encounter commonly elicit feelings of familiarity simply because of their resemblance to old objects. (2008, 11)

Farwell and colleagues, of course, agree with this assessment. They claim that Brain Fingerprinting is a technology that produces "no false negatives or false positives" (2012, 115). In short, while some think this technology could face problems in cases of forgetting, both proponents and critics of Brain Fingerprinting assume that this technology is immune to false positives because it is unlikely, if not impossible, for false recognition to occur.

9.4   False recognition

In the previous section I outlined the Archival View of Memory, to which both proponents and critics of Brain Fingerprinting are committed. Adopting this view has led to a collective oversight of the possibility of false recognition. This is a mistake. As I will show below, false recognition is not only possible, it is a prevalent and persistent form of memory error. What's worse, instances of false recognition show a P300 response indistinguishable from instances of genuine recognition.

9.4.1   The DRM paradigm: false recognition is possible

The Deese-Roediger-McDermott, or DRM, paradigm is one of the most prominent techniques by which false recognition is elicited experimentally (Deese 1959; Roediger & McDermott 1995). In the DRM, participants are presented with a list of related items, which they are asked to memorize. The most basic form of this task involves a list of semantically related words, such as *nurse, sick, medicine, lawyer, health, hospital,* etc. After a short break, participants are presented with additional items and asked whether they recognize them as members of the initial list. This recognition test involves three types of item: a) items from the list (*hospital*), b) items unrelated to and not presented on the list (*apple*), and c) items related to, but not presented on, the list (*doctor*).

Here's an example of a DRM experiment:

Figure 9.1   DRM paradigm

When asked whether they recognize *hospital*-type items, participant responses follow a standard serial position curve: the rate of recognition varies as a function of the item's location on the list. Participants are most likely to recognize words at the beginning or end of the list – doing so 70–80% of the time – but only recognize words in medial positions 40–50% of the time. Participants rarely report recognizing words of type b). When the word was not presented on and is unrelated to the previous list, as is the case with *apple* in the above example, participants report recognizing it less than 10% of the time. In the DRM task, the recognition questions of most interest are those of type c): words that are related to the presented items but were not on the previous list, such as *doctor*. Do participants treat *doctor* more like *hospital* or more like *apple*? Participant responses to *doctor* are similar to *hospital*; they report recognizing this non-presented word 70–80% of the time, a rate comparable to that of the best-recognized words from the initial list.

One might object to my characterization of *doctor*-type responses in the DRM task as instances of false recognition. After all, participants do not have to remember hearing *doctor* in order to answer the recognition question. Their false recognition may reflect a guess that *doctor* was likely to have been on the list, given that the list also featured words like *hospital* and *stethoscope*. While this interpretation is plausible given the results presented thus far, it is inconsistent with other measures taken during the DRM task. Each time a participant claims to recognize an item, he or she is asked to make an additional remember/know judgment about the response (Roediger & McDermott 1995; Lampinen, Neuschatz, & Payne 1999). Participants are asked whether they can recall specific, vivid details of the item's presentation, in which case they are instructed to respond that they remember the item. If participants are guessing when they report recognizing *doctor*, then they should report that they know doctor was on the list, rather than that they remember it. Responses for *doctor*-type items, however, are overwhelmingly remember-responses. Participants report remembering *doctor* as often as they report remembering items that were presented, such as *hospital*, 58% and 57% of the time, respectively, in Roediger and McDermott's (1995) initial study. In fact, participants are often willing to provide details about the presentation of this non-presented word, including where it occurred on the list, what they were thinking of as they heard it presented, and what the voice reading the word sounded like. For example, a participant in a Dewhurst and Farrand experiment claimed that the word *piano* – which was not an item on the list he learned – caused him to form an "image of trying to get a grand piano through the front door at home" (2004, 408).

9.4.2   False recognition is prevalent and persistent

Even if the DRM does provide evidence that false recognition is possible, one might still be inclined to downplay the significance of recognizing *doctor-* and *piano*-type items, dismissing this result as the product of a highly contrived and artificial task. But this interpretation, too, is difficult to maintain. First, the false recognition result in the DRM paradigm has been replicated extensively. According to a recent estimate, there has been "one new [DRM] experiment every 2 weeks for the past 10 years" (Gallo 2006, 22). The effect can also be obtained with a range of stimuli. False recognition occurs when the set of words are semantically related, as in the above example, but also when the words are related by category, phonology, or orthography.[6] So too, the effect can be found with non-linguistic items, such as pictures, faces, and dot arrays.[7]

The DRM effect is not only pervasive, it is persistent. In standard versions of the task, the learning and recognition events are separated by a few minutes, but the effect remains when the events are separated by hours, days – and in some cases – months.[8] Even more strikingly, the tendency to confidently report recognizing nonpresented items continues despite warnings to be vigilant against such errors, regardless of whether the warning is given during presentation of the initial list or at the time of recognition.[9] Third, the memory errors found in the DRM task are analogous to those found in memories for significant cultural events, such as the fall of the Berlin Wall, September 11, 2001, and the recent uprisings in Egypt. While these memories are often referred to as "flashbulb memories," in homage to their seeming indelibility, studies reveals that the details presented in these recollections change over time, even though participant confidence in their veridicality does not (e.g., Neisser & Harsch 1992). For example, a person might initially report having heard of the event while at work, only later to recall first seeing the event reported on television. As in the DRM task, participants are often confident in their reports of details that were not part of their previous experience, and in some cases retain this confidence even in light of evidence that tells against the accuracy of their retelling. Given the convergence between these results, many psychologists are optimistic that rigorous study of memory errors in restricted experimental contexts such as the DRM paradigm can be used to gain insight into the role of false recognition in more significant contexts, including eyewitness testimony (e.g., Loftus 2003).

9.4.3 False recognition is indistinguishable from true recognition

It appears that false recognition is a more prominent form of memory error than proponents of Brain Fingerprinting have acknowledged. This may not, however, present a problem for this method of interrogation. All of the measures of false recognition discussed thus far have relied upon behavior, revealing that participants respond similarly to list items and related lures. Since Brain Fingerprinting uses neural, rather than behavioral, indicators of recognition, it may be able to detect differences in how list items and lures are processed prior to the participant's overt response.

Unfortunately, ERP studies of the DRM effect have been unable to identify clear and consistent distinctions in the P300 response for truly and falsely recognized items. When differences are found, they have shown that the P300 response is *more* prominent for falsely recognized items than for ones that are truly recognized. In an initial test of ERP effects on the DRM Paradigm, Miller, Baratta, Wynveen, and Rosenfeld (2001) found no significant differences in P300 amplitude for list items and lures – words like *doctor* showed the same positive trend as words like *hospital*. They did find a difference in the P300 onset for these two stimulus types. For falsely recognized (*doctor*-type) words, the P300 began earlier than it did for truly recognized (*hospital*-type) words. Similarly, in a set of three experiments Allen and Mertens (2009) found only one case where the P300 response differed significantly for true and false recognition, and here again, the increase in positive deflection was stronger for false than true recognition. Recent studies have shown that this trend persists even when the initial event and recognition test occur in different modalities (e.g., the initial list of words is read aloud while the recognition test uses written words) (Beato, Boldini, & Cadavid 2012; Boldini, Beato, & Cadavid 2013). ERP studies of false recognition are ongoing, but there is general skepticism surrounding the idea that true and false recognition could be distinguished in the way Brain Fingerprinting requires:

> Differences between true and false recollections are not often found, and those that are uncovered are relatively subtle, and unlikely to be of use in classifying outcomes for individual subjects. (Allen & Mertens 2009, 487)

In response to the preponderance of false memory effects, many memory scientists have abandoned the Archival View of Memory. In its place they offer a Constructive View of Memory, according to which memory is nothing more than patterns of information, amalgamated across many past events, rather than traces of particular past events (Neisser 1967; Sutton 1998; Loftus 2003). As Schacter puts the point, "Memories are not stored as discrete traces but rather are superimposed on preexisting memories in a composite representation" (Schacter 1995, 19). Remembering is not an act of retrieval for the Constructivist. Instead, it is a matter of using the patterns stored in memory to construct a plausible representation of the event one wants to remember. Remembering the past, making decisions in the present, and planning for the future all rely on the same underlying cognitive processes; they are distinguished only by the tense assigned to the outcome (Schacter, Addis, & Buckner 2008).

9.5   Brain Fingerprinting and constructive memory

False recognition creates a challenge for Brain Fingerprinting and the view of memory on which it is based. How should our understanding and evaluation of this technology change as a result of the trend toward the Constructive View of Memory? I offer two suggestions below.

9.5.1   The false positive problem

At the very least, false recognition results show that false positives are a bigger problem for Brain Fingerprinting than previously supposed. Meegan (2008) and other critics of this technology have severely underestimated the likelihood of detecting an amplified P300 response to a novel stimulus. There is good reason, in fact, to think that false positives present a bigger challenge to Brain Fingerprinting than false negatives, contrary to what Meegan himself has claimed. People falsely recognize items for any number of reasons – the lure may have a similar shape to an item previously encountered, or it may be similar in color, function, or semantic category. Given the seemingly limitless ways in which the stimulus could resemble a previous event, it seems highly unlikely that a perpetrator could retain any information about the crime and fail to exhibit an amplified P300 response. Of course, cases of pure forgetting could still result in false negatives. But if memories are stored not as discrete traces, but rather patterns of information across events as the Constructive view suggests, then it seems that the forgetting would have to be extensive in order for information related to a crime to fail detection.

The situation is worse for Farwell and colleagues, who have claimed that there are no false positives in Brain Fingerprinting. This claim may be – technically – true in the experiments they report, but it does not rid Brain Fingerprinting of the worries brought on by false recognition. These experiments are not designed in a way that is sensitive to the false positives produced by false recognition. Recall that Brain Fingerprinting tests employ three types of stimuli: targets, irrelevants, and probes. When Farwell and colleagues report a lack of false positives, they mean there were no irrelevants that showed an amplified P300 and the probes showed this response only for participants who had been given the relevant information. They do not consider the possibility that someone could show an amplified P300 response without recognition. To do so, they would have to include an additional stimulus type, *lures*: items highly similar to probes, but not part of the previous encounter. To my knowledge, Brain Fingerprinting tests have never incorporated stimuli of this type.

The false positive problem will be difficult to remove from Brain Fingerprinting for two reasons. First, if the Constructive View of Memory is right, then false recognition is an endemic feature of memory. The purpose of memory, according to Constructivists, is to extract patterns from previous experiences, such that they can be used to guide future thought and action. Although false recognition is a memory *error*, it is not a *malfunction*. Much as seeing is susceptible to visual illusions of depth, motion, etc., memory falls prey to mnemonic illusions when presented with items that resemble, but do not replicate, previous experiences (Roediger 1996). False recognition occurs in experimental contexts where participants are trying to perform well on the task. It seems likely that these errors would be compounded in cases where the participants are suspects, many of whom may be motivated to conceal information from the interrogators.

Second, by failing to consider the possibility of falsely recognizing lures, Farwell and colleagues are ignoring the cases of false recognition that are most relevant in legal contexts. To provide evidence that could be used to establish guilt, Brain Fingerprinting would have to support subtle distinctions between descriptions of an event – whether three shots were fired or four, whether the getaway car was a *Camry* or a *Corolla*, whether the threat was explicit or implied. The possibility of false recognition is at its highest in such contexts, as these determinations involve scenarios that are highly similar to what actually occurred. The general human susceptibility to false recognition of similar stimuli is more than sufficient to supply reasonable doubt.

9.5.2 Interpreting the P300

False recognition creates difficulties for Brain Fingerprinting that go far beyond false positives, challenging the basic characterization of this approach to interrogation. Specifically, false recognition results reveal that the P300 response cannot be interpreted as a measure of recognition. An amplified P300 response is not an indication that the stimulus being presented matches a piece of information stored in the suspect's brain. This interpretation of the P300 is inconsistent with the well-replicated finding that instances of false recognition – where there is no corresponding information in the brain to be activated – produce a P300 response indistinguishable from that of true recognition.

Farwell and colleagues could try to relax their interpretation of the P300 somewhat, characterizing it as a measure of familiarity rather than recognition. Indeed, some neuroscientists have offered this interpretation (e.g., Rugg & Curran 2007). But this interpretation is equally hard to reconcile with the false recognition results. Recall that the few occasions on which a distinction between true and false recognition could be drawn were cases where the P300 response was *stronger* for false recognition cases than for true recognition. If the amount of P300 amplification is an indication of the degree to which a stimulus is familiar, then this would mean that participants treat lures as more familiar than items they actually encountered. This is, at best, counterintuitive.

The inability to characterize the P300 as a recognition response should not come as a complete surprise. After all, the amplified P300 response is found in a range of non-memory contexts, such as the Oddball Paradigm discussed in section 2.2, and tasks focused on attention. In addition, it is influenced by a variety of factors, from fatigue (Polich & Kok 1995) to intelligence (Polich 2007).

So what does the P300 measure? There are several competing explanations in contemporary neuroscience. Some consider it a signal of perceptual fluency (Voss & Federmeier 2011), whereas others view it as a measure of source monitoring (Beato et al. 2012), context updating (Donchin & Coles 1988), or attentional effort (Kramer & Strayer 1988). The question is not likely to be answered soon, if ever, as Polich explains in a recent review of P300 effects:

> Specifying a singular overarching explanation for this neuroelectric phenomenon [the P300] has proven difficult, primarily because the P300 is observed in any task that requires stimulus discrimination – a fundamental psychological event that determines many aspects of cognition. (Polich 2007, 2137)

If the P300 response is not an indication of recognition, then Farwell and colleagues lose their warrant for the claim that Brain Fingerprinting is a method of memory detection. It does not provide a way to "unlock the truth" that a suspect may have hidden in memory, as they have advertised.

With the loss of their recognition measure, Farwell and colleagues also lose their purported advantage over other methods of interrogation and lie detection. Brain Fingerprinting was supposed to be a superior technique because it detected information directly, rather than measuring physiological correlates of information or deception. It now appears that this characterization of Brain Fingerprinting was mistaken. The P300 is yet another correlation measure – a neurological marker that is sometimes associated with recognition.

This is not a criticism of the P300: it is unlikely that any other direct measure could be identified as its replacement. If the Constructive View of Memory is the right one, then recognition, strictly defined, no longer happens, as there are no traces of particular past events stored in memory. The act of remembering is always, at least to some degree, a process of construction. The search for any measure of pure memory detection may be a fool's errand.

Nothing I have said thus far rules out the possibility that Brain Fingerprinting is still the best of the available interrogation measures. It may be the most reliable indicator of a suspect's guilt or innocence, even if it fails to offer a foolproof way of unlocking memories. It's only that the difference between Brain Fingerprinting and the polygraph, or other forms of the GKT, is one of degree, not kind.

9.6   Conclusion

The advent of neural measures of interrogation and lie detection has piqued the interest of many working in law enforcement, as they suggest ways to circumvent suspect's attempts to conceal information during interrogation. With its promise to "unlock the truth within an individual's memory" (http://www.brainwavescience.com) Brain Fingerprinting has been considered especially intriguing.

I have offered an analysis of the claim that Brain Fingerprinting is a method of memory detection. As I have shown, this claim relies on a particular interpretation of the P300 response and, in turn, a particular, Archival View of Memory, neither of which can be supported. Brain Fingerprinting detects not only information and events that a suspect has seen before, but also information and events that are similar – in one respect or another – to things a suspect has previously encountered. There may still be a role for this technology in criminal investigations, but its limitations must be well understood.

Notes

1. Farwell characterizes the P300 MERMER as a positive peak followed by a late negative peak, or LNP (Farwell 2012).
2. The GKT is now sometimes referred to as the Concealed Information Test (CIT), which reflects how it used in counterterrorism (i.e., to determine whether the respondent has information about plans for crimes that have not yet been committed).
3. To ensure that suspects know these details, the investigators provide suspects with this information prior to the Brain Fingerprinting test.
4. Locke understood memory to be a power of the mind "to revive Perceptions, which it has once had, with this additional perception annexed to them, that it has had them before" (1690/1975: 150). Similarly, Hume questions, "For what is the memory but a faculty, by which we raise up the images of past perceptions?" (1739, Book 1, Part I, Section IV).
5. See Draaisma 2000 for further discussion of the metaphors used to describe memory and memory traces.
6. Category: Dewhurst & Anderson 1999. Phonology: Sommers & Lewis 1999. Orthography: Watson, Balota, & Roediger 2003.
7. Pictures: Koustaal 2006. Faces: Homa, Smith, Macak, Johovich, & Osorio 2001. Dot arrays: Nosofsky 1991.
8. Seamon, Luo, Kopecky, Price, Rothschild, Fung, & Schwartz 2002; Toglia, Neuschatz, & Goodwin 1999.
9. Anastasi, Rhodes, & Burns 2000; Gallo, Roberts, & Seamon 1997.

References

Allen, J. J. B. & Mertens, R. (2009). Limitations to the Detection of Deception: True and False Recollections Are Poorly Distinguished Using An Event-Related Potential Procedure. *Social Neuroscience*, 4, 473–490.
Anastasi, J. S., Rhodes, M. G., & Burns, M. C. (2000). Distinguishing between Memory Illusions and Actual Memories Using Phenomenological Measurements and Explicit Warnings. *American Journal of Psychology*, 113, 1–26.
Anderson, J. R. (1976). *Language, Memory, and Thought*. Hillsdale, NJ: Erlbaum.
Anderson, J. R., Bothell, D., LeBiere, C., & Matessa, M. (1998). An Integrated Theory of List Memory. *Journal of Memory and Language*, 38, 341–380.
Beato, M. S., Boldini, A., & Cadavid, S. (2012). False Memory and Level of Processing Effect: An Event-Related Potential Study. *Neuroreport*, 23, 804–808.
Ben-Shakhar, G. (2012). Current Research and Potential Applications of the Concealed Information Test: An Overview. *Frontiers in Psychology*, 3, 1–11.
Boldini, A., Beato, M. S., & Cadavid, S. (2013). Modality-Match Effect in False Recognition: An Event-Related Potential Study. *Neuroreport*, 24, 108–113.

Deese, J. (1959). Influence of Inter-Item Associative Strength upon Immediate Free Recall. *Journal of Experimental Psychology*, 58, 17–22.

Dewhurst, S. A. & Anderson, S. J. (1999). Effects of Exact and Category Repetition in True and False Recognition Memory. *Memory & Cognition*, 27, 664–673.

Dewhurst, S. & Farrand, P. (2004). Investigating the Phenomenological Characteristics of False Recognition for Categorized Words. *European Journal of Cognitive Psychology*, 16, 403–416.

Donchin, E. & Coles, M. G. (1988). Is the p300 Component a Manifestation of Context Updating? *Behavioral and Brain Sciences*, 11, 357–427.

Draaisma, D. (2000). *Metaphors of Memory: A History of Ideas about the Mind*. Cambridge: Cambridge University Press.

Farwell, L. A. (2012). Brain Fingerprinting: a Comprehensive Tutorial Review of Detection of Concealed Information with Event-Related Brain Potentials. *Cognitive Neurodynamics*, 6, 115–154.

Farwell, L. A. & Donchin, E. (1991). The Truth Will Out: Interrogative Polygraphy ("lie detection") with Event-related Potentials. *Psychophysiology*, 28, 531–547.

Farwell, L. A. & Richardson, D. C. (2006). Brain Fingerprinting in Laboratory Conditions. *Psychophysiology*, 43, S37–S38.

Farwell, L. A., Richardson, D. C., & Richardson, G. M. (2012). Brain Fingerprinting Field Studies Comparing p300-mermer and p300 Brainwave Responses in the Detection of Concealed Information. *Cognitive Neurodynamics*, 6, 115–154.

Farwell, L. A. & Smith, S. S. (2001). Using Brain mermer Testing to Detect Concealed Knowledge Despite Efforts to Conceal. *Journal of Forensic Sciences*, 46, 135–143.

Friedman, D. & Johnson, R. (2000). Event-Related Potential (erp) Studies of Memory Encoding and Retrieval: a Selective Review. *Microscopy Research and Technique*, 51, 6–28.

Fox, D. (2008). Brain Imaging and the Bill of Rights: Memory Detection Technologies and American Criminal Justice. *The American Journal of Bioethics*, 8, 34–36.

Gallo, D. A. (2006). *Associative Illusions of Memory: False Memory Research in DRM and Related Tasks*. New York: Taylor & Francis.

Gallo, D. A., Roberts, M. J., & Seamon, J. G. (1997). Remembering Words Not Presented in Lists: Can We Avoid False Memories? *Psychonomic Bulletin and Review*, 4, 271–276.

Harrington v. State (2001). Case No. PCCV 073247. Iowa District Court for Pottawattamie County, March 5, 2001.

Homa, D., Smith, C., Macak, C., Johovich, J., & Osorio, D. (2001). Recognition of Facial Prototypes: the Importance of Categorical Structure and Degree of Learning. *Journal of Memory & Language*, 44, 443–474.

Hume, D. (1739/1978). *A Treatise of Human Nature*, L. A. Selby-Bigge (ed.). Oxford: Clarendon Press.

Iacono, W. G. (2008). The Forensic Application of "Brain Fingerprinting:" Why Scientists Should Encourage the Use of P300 Memory Detection Methods. *American Journal of Bioethics*, 8, 30–32.

James, W. (1890). *The Principles of Psychology*. London: Macmillan.

Koustaal, W. (2006). Flexible Remembering. *Psychonomic Bulletin & Review*, 13, 84–91.

Kramer, A. F. & Strayer, D. L. (1988). Assessing the Development of Automatic Processing: An Application of Dual-Track and Event-Related Brain Potential Methodologies. *Biological Psychology*, 26, 231–267.

Lampinen, J. M., Neuschatz, J. S., & Payne, D. G. (1999). Source Attributions and False Memories: a Test of the Demand Characteristics Account. *Psychonomic Bulletin & Review*, 6, 130–135.

Locke, J. (1690/1979). *Essay Concerning Human Understanding*, ed. P. H. Nidditch. Oxford: Clarendon.

Loftus, E. F. (2003). Our Changeable Memories: Legal and Practical Implications. *Nature Reviews: Neuroscience*, 4, 231–234.

Lubow, R. E. & Fein, O. (1996). Pupilary Size in Response to a Visual Guilty Knowledge Test: New Technique for the Detection of Deception. *Journal of Experimental Psychology: Applications*, 2, 164–177.

Lykken, D. T. (1959). The gsr in the Detection of Guilt. *Journal of Applied Psychology*, 43, 385–388.

Meegan, D. V. (2008). Neuroimaging Techniques for Memory Detection: Scientific, Ethical, and Legal Issues. *The American Journal of Bioethics*, 8, 9–20.

Miller, A. R., Baratta, C., Wynveen, C., & Rosenfeld, J. P. (2001). p300 Latency, But Not Amplitude Or Topography, Distinguished between True and False Recognition. *Journal of Experimental Psychology: Learning, Memory and Cognition*, 27, 354–361.

Neisser, U. (1967). *Cognitive Psychology*. New York: Appleton-Century-Crofts.

Neisser, U. & Harsch, N. (1992). Phantom Flashbulbs: False Recollections of Hearing the News about Challenger. In E. Winograd & U. Neisser (eds), *Affect and Accuracy in Recall: Studies of Flashbulb Memories* (pp. 9–31). Cambridge: Cambridge University Press.

Niedermeyer E. & da Silva F. L. (2004). *Electroencephalography: Basic Principles, Clinical Applications, and Related Fields*. Philadelphia: Lippincott, Williams, & Wilkins.

Nosofsky, R. M. (1991). Tests of An Exemplar Model for Relating Perceptual Classification and Recognition Memory. *Journal of Experimental Psychology: Human Perception & Performance*, 17, 3–27.

Polich, J. (2007). Updating p300: An Integrative Theory Of P3a and P3b. *Clinical Neurophysiology*, 118, 2128–2148.

Polich, J. & Kok, A. (1995). Cognitive and Biological Determinants of p300: An Integrative Review. *Biological Psychology*, 41, 103–146.

Proust, M. (1913). *Remembrance of Things Past. Volume 1: Swann's Way: Within a Budding Grove*, trans. C. K. Scott Moncrieff and Terence Kilmartin. New York: Vintage.

Roediger, H. L. (1996). Memory Illusions. *Journal of Memory and Language*, 35, 76–100.

Roediger, H. L. (2000). Why Retrieval Is the Key Process to Understanding Human Memory. In E. Tulving (Ed.), *Memory, Consciousness and the Brain: the Tallinn Conference* (pp. 52–75). Philadelphia: Psychology Press.

Roediger, H. L. & McDermott, K. B. (1995). Creating False Memories: Remembering Words That Were Not Presented in Lists. *Journal of Experimental Psychology: Learning, Memory, and Cognition*, 21, 803–814.

Rugg, M. D. & Curran, T. (2007). Event-Related Potentials and Recognition Memory. *Trends in Cognitive Science*, 11, 251–257.

Schacter, D. L. (1995). Implicit Memory: A New Frontier for Cognitive Neuroscience. In M. S. Gazzaniga (ed.), *The Cognitive Neurosciences*(pp. 815–824). Cambridge, Mass.: MIT Press.

Schacter, D. L., Addis, D. R., & Buckner, R. L. (2008). Episodic Simulation of Future Events: Concepts, Data, and Applications. *Annals of the New York Academy of Sciences*, 1124, 39–60.

Seamon, J. G., Luo, C. R., Kopecky, J. J., Price, C. A., Rothschild, L., Fung, N. S., & Schwartz, M. A. (2002). Are False Memories More Difficult to Forget Than Accurate Memories? the Effect of Retention Interval on Recall and Recognition. *Memory & Cognition*, 30, 1054–1064.

Senkfor, A. J. & Van Petten, C. (1998). Who Said What? An Event-Related Potential Investigation of Source and Item Memory. *Journal of Experimental Psychology: Learning, Memory, & Cognition*, 24, 1005–1025.

Sommers, M. S. & Lewis, B. P. (2000). Who Really Lives Next Door: Creating False Memories with Phonological Neighbors. *Journal of Memory and Language*, 40, 83–108.

Sorabji, R. (2003). *Aristotle on Memory* (2nd edn). London: Duckworth.

Spence, S. A., Farrow, T., Herford, A. E., Wilkinson, I. D., Zheng, Y., & Woodruff, P. W. (2001). Behavioral and Functional Correlates of Deception in Humans. *Neuroreport*, 12, 2849–2852.

Sutton, J. (1998). *Philosophy and Memory Traces: Descartes to Connectionism*. Cambridge: Cambridge University Press.

Toglia, M. P., Neuschatz, J. S., & Goodwin, K. W. (1999). Recall Accuracy and Illusory Memories: When More Is Less. *Memory*, 7, 233–256.

Verleger, R. & Berg, P. (1991). The Waltzing Oddball. *Psychophysiology*, 28, 468–477.

Voss J. L. & Federmeier, K. D. (2011). fn400 Potentials Are Functionally Identical to n400 Potentials and Reflect Semantic Processing during Recognition Testing. *Psychophysiology*, 48, 532–546.

Watson, J. M., Balota, D. A., & Roediger, H. L. (2003). Creating False Memories with Hybrid Lists of Semantic and Phonological Associates: Over-Additive False Memories Produced by Converging Associative Networks. *Journal of Memory and Language*, 49, 95–118.

Part III
Evaluation and Speculation

10
Feminist Approaches to Neurocultures

Sigrid Schmitz

Results of brain research today are of high relevance for explaining and predicting individual behavior, cognition, and decision-making. The knowledge production and the scientific framework that is elaborated in the field of neurosciences, particularly by using brain imaging studies to locate the structural and functional fundament for thinking, feeling, and acting within the brain's matter, is taken evidently in a wide scope of other disciplines and contemporary academic discourses. With the expansion of the domains of neuroscientific knowledge, an abundance of *neurocultures* emerges. Not only the sciences and psychology, but also education, economics, sociology, the humanities, and philosophy take reference to the results of brain research and account for the individual as a "cerebral subject" (Ortega & Vidal 2007) – the cultural figure of the human according to which the self is constituted by its brain.

Although references to brain basics for explaining individual behavior or personhood, and even for legitimizing social order, have a long history (Hagner 2006), there seems to emerge a renewed concept of bio-essentialism in contemporary western society by using "the neuro" in approaches of neuro-pedagogy, neuro-economy, neuro-marketing, neuro-theology, neuro-aesthetics, amongst others. Particularly the setting of this prefix 'neuro-X' implies a *concept of genealogy*, wherein neurophysiological settings are conceptualized not only as being the ground source but also as providing the comprehensive explanatory value for behavioral developments, identity, and selfhood. In such, the notion of the cerebral subject sets the biological in paramount order to the social, it prior-ranks nature to culture.

The most interesting question, however, may be not the naturalization per se, but its current re-addressing in line with neoliberal concepts of contemporary Western culture that highlights individualization,

195

self-responsibility, flexibility, and rationalized optimization of their own performance for self-marketing in modern meritocracy (Pitts-Taylor 2010, Schmitz 2012). The biologically determined brain superficially is contrary to the notion of 'feasibility rather than fate' according to which bodies and brains should be shaped along the brain's fundamental modification and manipulation capacity up to theories of optimization. As bodies and brains should be used by the self and others as malleable resources that are applicable according to permanently changing societal settings, it has to be questioned: To which notions of normativity and power relations do the invocation of individual agency and responsibility for shaping the own brain, i.e., the personal biological matter, refer to? This is the question about the relation between changing and resistant norms according to the micro- and macro-space of modern biopolitics. In parallel, however, it is also the question about the normative dimensions of neuroscience (Pickersgill 2013): How far does the localization of preferences, skills, traits, personalities, and subjectivity within a concept of "brainhood" (Vidal 2009) serve normative attributions to particular groups and their goals, and how does it legitimize in- and exclusions in social spheres?

In this paper I will particularly focus on gender norms that are re-negotiated within neuroscientific research and neurocultures. The aim is to elaborate the existing scientific controversies about sex/gender[1]-related brain research and the impacts of current neurocultures on gender based in- and exclusions within socio-cultural power relations. Instead of rejecting the 'neuro', neurofeminism (Bluhm et al. 2012) takes a bio-cultural view as the background to account for the inseparable entanglements between the developing biological matter and social influences, i.e., the "ineluctably social character of nature and the natural makeup of the social" (Pitts-Taylor 2011).

By using feminist materialist approaches (cf. Alaimo & Hekman 2008) the potentials of re-defining gender aspects in brain research and neurocultures can be analyzed as well as the impacts of current understandings of a neurologically framed gender, i.e., the transmission of seemingly uncontroversial neuroscientific facts into socio-cultural discourses and gendered power relations that may justify, rather than question, discrimination and social inequality along gender and intersected axes.

In the following I firstly present a short review of neurofeminist approaches and critical evaluations on sex/gender brain research that enable a differentiated view on the theoretical, methodological, and empirical background, and on the interpretations particularly drawn from neuroimaging studies with respect to sex/gender differences. Secondly, I focus on the concept of brain plasticity aiming to discuss its potentials for deconstructing sex-based neurodeterminisms and instead to highlight the mutual entanglements of the biological and the social in brain-behavior development. I will also suggest how this concept has to be further developed under the framework of feminist materialism to account for the social construction of the gendered brain as well as for the brain's agency within these networks. Based on this neurofeminist background, I will thirdly question the enclosure of plasticity concepts in a 'modern' form of neurobiological determinism that – in my view – frames the perspectives of current neurocultures and may explain the synchrony of both the essentialist notions of brainhood and the governmental demands of self-responsible brain-improvement for particular goals in neoliberal western society. Within the two neurocultural fields of neuroenhancement and social neuroscience, I work out the nevertheless resisting gender norms in the framing of the malleable cerebral subject that especially refer to masculinized connotations of rationality in contrast to feminized connotations of emotionality. I conclude in questioning why these gendered norms remain so deeply embedded in neuroscientific and neurocultural discourses and how approaches of neurofeminism and feminist materialism can help to support a differentiated view.

10.1 Neurofeminist approaches to sex/gender brain research

Critical analyses of the methodological and empirical background of brain research uncovered its strong focus on sex difference research. The cultural notion of two sexes often leads to a difference-oriented methodology, through which each group is assumed to be inherently homogenous. Current scientific data, however, suggest a far more complex picture than the reference to 'brain differences' assumes. The scope of neuroscientific findings in terms of cognitive abilities (e.g., linguistic ability, spatial orientation, or mathematics), processing of emotional features, or brain basics concerning sexual orientation (including hormonal interactions) are by no means conclusive (for overview Bluhm et al. 2012, Kuria 2012, Jordan-Young 2010, Jordan-Young & Rumiati 2012, Roy 2012, Sommer et al. 2004, Vidal 2012). Differences within the groups of women and men often transgress the boundaries of gender, and more similarities between the gender groups can be found on the other hand (Hyde 2005). Nevertheless, the highlighting of differences instead of researching similarities or intra-gender group variations remains to be the primary focus of sex/gender brain research. The publication bias to a higher amount of citations of difference-studies than of studies that did not find sex/gender differences remains a common practice in neuroscientific research and even more in the transgression of its results into popular media (cf. Fine 2010).

A particular focus of critical neurofeminist research is laid on the impacts of the experimental design and procedures of data analyses on the drawing of conclusions regarding the statements of sex differences within specific domains of brains structure, function or structure-function relationships. Methodological variations in data selection, statistical analysis, and computer tomographic calculations (cf. Haller & Bartsch 2009, Kaiser et al. 2009, Schmitz 2010, Vul et al. 2009) speak against inaccurate generalizations of results concerning a binary two-sex categorization. The acceptance of first reviews on methodical choices that influence neuroscientific results (cf. Kaiser et al. 2009, Wallentin 2009) in scholarly journals attests an increasing sensitization towards a critical reflection of methods within the neuroscientific community.

Another focus is set on the constructive processes in brain imaging procedure per se that aims at localizing and visualizing brain structures and functions at time and in line with information processing and action with historically new methods, such as functional magnetic resonance imaging (fMRI). No question, brain imaging has turned out to be an important technique to improve the knowledge of brain processes. However, brains in neuroimaging procedures experience a double transformation en route. The body itself stays behind in the scanner, its data are processed and converted into colorful points on the screen, and in the end of a research these data are presented as images or atlases of the 'real' brain. This transformation from corporal matter to non-corporal data and conversion back to an image of a brain's matter is certainly not unprecedented. It has existed from the beginnings of illustrations, to photography, to the current digitalization techniques of computer tomography or fMRI, and it has always included conscious or unconscious decisions by the researcher of what should appear in the image during reproduction and what should not appear, or how it should appear.

Approaches from Science and Technology Studies have outlined specific "inscriptions" (Latour & Woolgar 1979) that go in line with these constructive processes of brain imaging. Situated in the theoretical contexts in which the images are produced, they depend on the processes of negotiation within a lab, the aims of a research, the conscious and unconscious understandings and beliefs of researchers, to name only a few intervening factors (for detailed overview, see Beaulieu 2001, 2002). This does not mean that knowledge about the brain, which has been gained through imaging technology, is random or inapplicable. On the contrary, especially within the medical field, ways of constructing knowledge through specialized procedures are useful for different diagnostic fields, therapy or neurosurgery.

However, brain images become problematic when they are used out of the context in which they were developed to make over-generalized statements about predefined groups based on gender, ethnicity, or other universalized categories. Gender assumptions, cultural notions, and norms are inscribed within the selection of particular methods during an experiment out of a broad range of scientific possibilities, e.g., applied algorithms, computations, statistical analyses, and computer graphical methods (cf. Fitsch 2012, Kaiser et al. 2007, Nikoleyczik 2012). For example, Kaiser et al. (2007) have shown how the use of different thresholds (all part of the scientifically accepted scope in data analysis) that separate task-oriented activations from background noises result in different images of the same data. Dependent to the choice of the threshold that had been taken for data analysis, the images of average activation in Brocas area during language tasks from 22 females and 22 males presented group differences or non-differences. This dependency on methodology is gaining importance because various research groups use different methods in image construction and data analysis, and there is hardly a study that is perfectly comparable to another.

10.2 Potentials and challenges of brain plasticity

An important step of critical neurofeminist research is to question the theoretical underpinnings of the empirical 'evidence' and to challenge the interpretation of findings. The concept *brain plasticity* points out that brain structures and brain functions are not in any shape or form determined by evolution or remain unchanged during a life span. At birth, the brain is not at all branded or defined, and this network of nerve cells, neuronal fibers and their synapses is not 'completely formed' by genetic information. As Catherine Vidal puts it: "The human brain is made up of 100 billion neurons and 1 million billion synapses which are the junctions between neurons, while there are only 6000 genes involved in the nervous system. This means that there are not enough genes to control the building of our billions of synapses" (Vidal 2012, 297).

Within neuroscientific research, plasticity concepts were already formulated and worked out from the 1970s onwards. Findings from animal studies led to the conclusion that the processing of environ-mental stimuli already prenatally effects the development, decomposi-tion, and alteration of synapses between neurons. Learning experience stabilizes and destabilizes the central nervous network along its situat-edness in a particular environmental context to which it is supposed (Hubel & Wiesel 1970). This concept of brain plasticity and the brain's ability to adapt to its environmental influences applies to all areas of the brain, but outstanding to the most complex networks of the cortex. As a result, neuronal networks 'learn' repetitive patterns of information and *embody* them structurally and functionally. Neuroplasticity studies in the 1980s and 1990s have pointed to the adaptive potential of sensory and motor areas in primates (e.g., Kaas et al. 1990, Jenkins & Merzenich 1987) and in the last decades plasticity in the human cortex is in the center of a whole scope of analyses (see below).

Environmentally influenced neural and synaptic plasticity, down to physiological regulations and gene expression (for an overview, see Kandel et al. 2000), are taken as the cellular and molecular basics for learning principles in the central nervous system. The alteration of neuronal networks is a lifelong process, which occurs through experi-ence and with every learning activity. Brain plasticity is not only the basis for learning over the entire lifetime, but is also necessary for brain functions. The tremendous dynamic 'nature' of brain plasticity, particu-larly in humans, this constant interplay between the outside environ-ment and biological structure, is considered as a decisive advantage for human cognitive abilities.

Plasticity concepts have also been taken up in neurofunctional imaging research (cf. Schmitz 2010, Vidal 2012). Brain imaging studies have outlined the development of differentiated language networks according to language biographies (Bloch et al. 2009), changes in hippocampal synaptic density in relation to navigational experience (Maguire et al. 2000), or even short-term effects of a bi-manual three-month training on neuronal density and interconnectivity in cortical motor areas (Draganski et al. 2004).

I deliberately worked out this history of neuroscientific research on brain plasticity to point to the fact that this concept has been discussed within the scientific community now for about 40 years. This notion is impor-tant insofar as plasticity discourse thus should also impact on neuroscien-tific research concerning sex/gender differences. One would expect that the research on environmental influences, learning experiences, and the impacts on brain-behavior development within social and cultural situat-edness should be a designated part of neuroscientific sex/gender research.

10.2.1 Plasticity in sex/gender research

From a neurofeminist perspective the materialization of the brain, i.e., the development of biological structure, physiological processes, and activation networks, turns out to be embedded in a constructive process between nature and culture. The networks of nerve cells and synapses that process information and regulate behavior need input from the world outside of the self, the world that the subject perceives and in which it behaves. The plasticity concept can help to explain inter-individual variation, i.e., differences between individuals as well as intra-individual variability, the changing of everybody's brain throughout life span (Schmitz 2010).

If the variability of brains is based on the *embodiment* of mental, social, societal, and cultural experiences in its individual structure and function, the sex/gender features according to the brain should likewise constitute through an ongoing dynamic process of bio-psycho-socio-cultural interactions. In following plasticity approaches, the neurofeminist discourse questions the bi-polarization of determined sex *against* constructed gender differences in the brain and instead emphasizes the inseparability of biological processes and social influences. Detailed research on the development of gender role and gender identity currently worked out such complex entanglements (Fausto-Sterling et al. 2012a, 2012b). Additionally, findings on the behavioral level provide hints that the perception of gender stereotyped cognitive skills influence the own cognitive performance more than biological sex differences (cf. Fine 2010).

The brain's capacity to adapt according to experiences also challenges the interpretative value of brain images. How could neurofunctional images describe an organ that is itself in constant change? Every brain image and every brain atlas is only a *snap-shot* (Schmitz 2010) of a current state in individual development, in a way, a status quo only at a specific point in time. The presentation of a particular structure or activity pattern (even if leaving the constructive 'nature' of brain images aside for a moment) always includes biological development and individual biography. The determination of cause and effect in this snapshot is pure interpretation if not speculation. Neither can the image be taken as a proof of biological determination of this state (sex), nor does it provide direct evidence for the incorporation of learning influences, societal stereotypes, or cultural norms (gendered construction of brain matter). The brain image as an empirical finding is per se 'underdetermined' for legitimizing or contradicting one or the other hypotheses. This crucial aspect of underdetermination of empirical results for scientific theories (including sex/gender assumptions) has been stressed by feminist science studies for a long time (Harding 1987).

Cordelia Fine (Fine 2013) recently analyzed the scope of neurofunctional imaging studies within two years (2009–2010) that refer to sex/gender differences. In the light of the ambiguous findings (besides methodological biases), she points to the following: "Similarly, researchers could, on the basis of ambiguous neurological findings, frequently speculate about the psychological effects of socialization or context on gendered behavior in a piecemeal fashion that is not developed into explicit neurocognitive models and systematically tested" (Fine 2013, 399). Her analyses, however, uncovered that none of the 39 studies she had identified took such a 'plasticity' approach; instead all of them followed the snap-shot approach (ibid., 397). Thus, although the brain plasticity concept is an outstanding perspective of knowledge production within the field of neuroscience, it seems to vanish out of view constantly when it comes to sex/gender research.

On the one hand, this may be due to the methodological focus of brain research to functional imaging technology which limits the analysis of plasticity processes according to its standardized technology (as mentioned above). Moreover, however, it seems that according to the question of sex/gender aspects in the brain, researchers often position themselves into different schools that are hardly in dialogue with each other. The old barricades between 'nature' and 'culture' still seem insurmountable.

10.2.2   How to bridge the gap of nature/culture and sex/gender

Along the aim to speak against biologist determinations of sex differences, neurofeminist discourse has stressed the socio-cultural conditions that impact on the structure and the function of the brain – owing to the brain's plasticity that is constantly open to the environment. In following the constructivist perspective of feminist discourse (cf. Butler 1993) this view is important to account for social power relations that performatively construct not only identity and subjectivity but also the corporal manifestation of gendered norms and values. The call for "[q]ueering the brain" (Dussauge & Kaiser 2012) would then imply inverting the question of cause and effect, to account for the constitution of the brain through society, to emphasize the materialization of experience in bodily matter. In referencing these post-structuralist approaches, Dussauge and Kaiser call for a re-formulation of brain theories. There is no 'nature' outside 'culture,' and thus, the gendered brain can only be referred to within the framing and re-cognition of meaning-making processes, power relations, and cultural norms.

On the other side, post-structuralist approaches tend to disregard the question of matter, and the current discourse on *feminist materialism* tries to re-account for the active role of matter, for its agency in the development of worldly phenomena. Karen Barad conceptualized this approach as "agential realism" (Barad 2003). It is important to notice that based on the perspective of feminist materialism the agency of matter (including biological and non-biological agents) is not understood in forms of conscious intentionality but as a form of enactment that takes part in the realization of worldly phenomena. Phenomena (not things) come into being via mutual intra-actions between organic, technical, *and* semiotic practices. Consequently, the co-construction of matter and meaning is addressed within this onto-epistemological framework (Barad 2007). The conceptualizations of feminist materialism are not as new as they are sometimes presented. There are several authors who have pointed to the inseparability of material and linguistic aspects and to the co-construction of nature and culture, particularly accounting the sex/gender debate, that has been addressed by feminist science studies for a long time (for overview, see Dolphijn & van der Tuin 2012). Although there is a debate on the question of the newness of feminist materialism and whether this approach shifts or re-writes post-structural theories into a new form of ontology (cf. van der Tuin 2011), I state that what is new in this approach is the more detailed focus on the intra-active processes between material becomings and meaning-making processes that are now taken into feminist research (cf. Alaimo & Hekman 2008).

Consequently, for a neurofeminist epistemology referring to brain plasticity this means to conceptualize the brain-in-the-world not as a passive foil awaiting the inscriptions of norms and values into its structures and functions, but to account for the inseparable co-construction of both brains' agency and meaning-making in the processes of brain-behavior-becomings.

The attempt to research the mutual intra-actions of the brain's dynamic materiality with the emergence of thinking, feeling, and behavior within normative socio-cultural discursive practices in a non-essentialist manner has already been addressed by Elisabeth Wilson, who grounded her arguments in theories of neuro-connectionism (Wilson 1998). Recently, these attempts have been taken up to evaluate the various meanings of brain plasticity in current neurocultural developments and discourses (Pitts-Taylor 2010, Schmitz forthcoming). I take the plasticity discourse as the starting point to step in again to the concept of brainhood and the setup of normative demands for the cerebral subject in a neoliberal society.

10.3   Plasticity, neoliberalism, and the gendered cerebral subject

As I have worked out above, the dynamic processes of plastic brain development speak for a continuous transgression of the border of 'nature' and 'culture'. At the same time new forms of neuro-governmentality (Schaper-Rinkel 2012) call for malleable brains to be optimized for enhancing personal competencies, for sharpening concentration and flexibility, and for improving mood and self-confidence by the use of diverse technologies, both from the self and from outside, which intervene in the body on a deep level. These discourses can be situated at the interface of medical, economic, and even philosophical literature referring to notions of individual responsibility for self-marketing and successful positioning in meritocracy. Although the use or rejection of brain optimization is shaped as an autonomous decision for promoting subjective agency, these decisions are embedded in a network of norms and values on the capitalist labor market.[2]

A note has to be considered to the – at first sight – contradictory discourse in neurocultures that, on the one hand, refers to the brain's biological essence in explaining and predicting cognition, behavior, and the self. On the other hand, the figure of the cerebral subject also includes the concept of brain plasticity in framing the biological matter as being open for a mutual form of shaping along particular goals. How can these two notions be made compatible? I have argued elsewhere that the concept of *modern neurobiological determinism* conceptualizes the brain in its current state as predicting and determining all thinking and behavior, independently, whether its structures or functions are inborn or learned (Schmitz 2012). In consequence, this concept can be also described as a snap-shot approach. At a particular point of acting or measuring, the 'biography' of the brain is seen as a result of bio-cultural development. But nevertheless, at this point the current structure and function of the brain is taken as the explanatory essence for all forms of acting. Thus, it is not the brain that is determined per se, but the brain should determine all outcomes of the individual's life.

It is this notion of neurobiological determinism that sets the ground for the figure of the cerebral subject in neoliberal contexts. 'We do not only have our brains, we are our brains', and this paradigm of "bio-identities" (Ortega & Vidal 2007, 257) in parallel provides the connection to new forms of current politics of a "biological citizenship" (Rose & Novas 2005). The neuronal self is responsible for self-discipline and self-technologies to promote health, fitness, flexibility, and success in the project-oriented polis of contemporary capitalism (Boltanski & Chiapello 2005). At the same time the closed and essential materiality that can be formed as a sort of material resource seems to legitimize these forms of neuro-governmentality. Just based on this argumentation the manipulation and target-oriented optimization of brains seems to be possible in a controlled manner. And only then – without disturbing forwardness – brains can be used as profitable resource on the neoliberal market. In my view, this framing of the brain-as-resource stands in contrast to the framework of the intra-active phenomenon of the brain-in-the-world, the former again separating matter and meaning, whereas the latter would ask in more detail how the entanglements between brain and self-technological practices realize with which outcomes.

In any case, the framing of the cerebral subject as the 'neuronal self' or the "neuro-chemical self" as Nicolas Rose named it (Rose 2007) links the fields of neuroscientific research and application to societal and public discourses. The new 'transdisciplinary' fields of neuro-pedagogy, neuro-economy, social neurosciences, or neuroenhancement are based on neuroscientific knowledge production. In particular, seemingly brain imaging 'facts' (I have presented the constructive and decisive 'nature' of these image-based facts above) provide a model for the localization of the bio-material source of the self. In the medical field disorders of even this cerebral subject are aimed to be mended with the help of technological interventions, or at least, related disabilities should be diminished (enabling). Further, manipulation and optimization with the help of neuro-technological applications shall also improve skills and performances for the so-called healthy subject.

The notion of an authorized neuroscientific knowledge production, that guides the emerging neurocultures to a great extent, points to the need of reassessing and addressing the impacts of sex/gender brain research on these discourses and their related practices. This holds well for the scientific fields and for the public understanding in order to gain particular insight into the transformation or persistence of gendered norms and values that accompany the mutual entanglements between brain research and the social, political, and public spheres.

In many neurocultural discourses (e.g., neuro-pedagogy, social neurosciences, or neuro-economics), the cerebral subject seems to be gender-neutral at first sight. Due to the notion of the malleable brain, biological determination may lose importance in legitimizing social positioning by seemingly sex-based inequalities, and at first glance it may also lose its power for legitimizing gender-related segregation and oppression in social and political contexts. However, in a recent analysis I have worked out how psychological and behavioral characteristics predicted by neuroscientific research are still interpreted and hierarchized in relation to particular gender norms; that is in particular the hierarchical division between masculine-connoted rationality and feminine-connoted emotionality (Schmitz 2012). In the following remarks I further elaborate on some aspects of these connotations in two fields of neurocultural discourses that seem to resist deconstructive attempts.

10.3.1   Gendering the field of neuroenhancement

For the rationally planned and designed increase of mental abilities and capabilities self-techniques of neuroenhancement are widely discussed to reach maximum performances in various societal fields. With the help of neuropharmaceuticals, such as Modafinil or Ritalin, memory capacities and concentration should be improved, or anti-depressives, such as Prozac, should regulate the mood to improve self-confidence.[3] This idea of improving the self by 'brain-tuning' is based on an understanding of effectiveness, which purports an observable indication of initiative, adaptability, and dynamism. Moreover, in this regard, individual achievement potential is no category of chance, but a project of the subject's own responsibility due to her/his ability to act as a 'rational' agent.

Since a decade now, science and the public intensively debate the use of neuro-enhancers. In the first half of the 2000s scientists and neuroethicists discussed the use of pharmacological enhancers mostly critically, arguing that their medical effects and social consequences were ambiguous. Central threads of this discourse concern medical impacts and their assessment (What side effects or long-term consequences are to be feared?), raise questions of equal opportunities and equal access to justice (Who do they benefit and how?) as well as identity (Do or don't neuro-enhancers change the personality?) (cf. Farah et al. 2004). Whereas in 2007 an article was published in the leading journal *Nature* that already presented a different tenor of the academic community by questioning the legitimacy to use 'professor's little helpers', if medical side effects turn out to be low (Sahakian & Morein-Zamir 2007). On October 10, 2008, another *Nature* article even argued a step further in legitimating the 'responsible' use of neuro-enhancers with the statement of evolutionary possibility and following an obligation to optimize the own brain, not only for personal success but also for the enhancement of a whole community or state (Greely et al. 2008). Although these examples cover only some facets of the widespread discourse, it is remarkable how fast this debate has changed.

In terms of impact assessment, proponents of neuroenhancement argue that such a 'brain-tuning' would be harmless due to its short-term effects and therefore similar to caffeine or other agents for increasing concentration (cf. Hall 2004). However, the long lasting effects of these pharmaceuticals on brain plasticity had been in the center of neuro-medical research and treatment (Normann & Berger 2008). Neuropharmaceuticals, such as Modafinil, Ritalin, or Prozac, had particularly been developed to regulate (stimulate or inhibit) synaptic efficiency and transmitter release and uptake. In such, the application of neuropharmaceuticals aimed at inducing structural-functional changes in the neuronal networks; some agents even influence the activity of neuronal genes which likewise impact synaptic connectivity (Angel et al. 2003). Although the neuroscientific knowledge only spoke against short-term effects but for sustainable impacts of neuroenhancement on the brain's composition, these consequences seem to be hazarded in view of the benefit for personal academic success.[4] It becomes clear in this cross-view that the arguments of medical assessment seem to vanish more and more in favor of economic prospects. Moreover, although proponents stress the short-term use of neuro-enhancers to increase a specific capacity at a particular time, the constant demands of the performance-oriented society are centered on a recurrent, thus sustainable capability. Under this constant pressure a possible addiction concerns not only physical but also psychological effects.

If neuroenhancement is conceived as a self-technology for upgrading rational skills and for the improvement of self-confidence and emotional stability, the question arises, how these discourses are related to gendered ascriptions to rationality and emotionality. The associations of Ritalin with the improvement of male users' concentration (coming from ADHD therapy with a higher degree of prescriptions to boys, Singh 2002) in contrast to the linking of the anti-depressive Prozac to female users (Blum & Stracuzzi 2004) have laid the ground for such gendered ascriptions. Although de-gendering arguments have been promoted in a way that the uptake of these pharmaceuticals could equalize the gender gap in behavioral characteristics, sometimes valued as positive (Kramer 1994), sometimes as negative (Fukuyama 2003) outcomes on a societal level, the gendered setup of neuro-enhancers seem to show a lasting persistence (cf. Schmitz 2012).

In order to shed light on the current popular discourse according to these gendered associations in neuroenhancement, a colleague and I have analyzed some argumentative assumptions circulating within popular media in Germany (Höppner & Schmitz 2014). The analysis of

21 articles that were published in four German online journals between 2006 and 2011 showed that self-optimization of the brain with the help of neuropharmaceuticals is today increasingly predicted as a strategy of success for everybody. However, repeated references to the user groups of particular neuro-enhancers show a similar picture as outlined above. Whereas success-oriented males are framed to aim to improve rational skills, success-oriented-women are framed to set preference on regulating their self-confidence.

Moreover, the articles suggest a hierarchized status quo of rational skills of non-enhanced women and men and a different proficiency level that users of neuro-enhancers can achieve according to their gender: While women would need a continual consumption of neuro-enhancer to achieve a proficiency level similar to 'male' capacities for a limited time scale, men should take neuro-enhancers only once in a while to improve selectively their apparent high potential (due to biological characteristics) in order to become the best within the group of the best men. In contrast, a continual consumption of Ritalin by women is shaped as resulting in distress and negative reactions of the social environment.

Outstanding in the current popular debate is the paradigm of individualization and self-responsibility in a neoliberal society. Equal opportunities qua optimization of the brain are addressed in the sense of 'everyone is his/her own fortune'. This goes in line with a disregard of social structures that perpetuate gendered discriminations on the labor market. Central to this logic is the individualization of a 'weaknesses of personal skills or traits' that needs improving, instead of fighting societal causes of these inequalities. Thus, the possibility of achieving equality only appears as being neutral. Who can afford neuro-enhancers? Access barriers are appreciable according to social class, ethnicity, and gender as well as economic situation. With the argument that society simply is unjust, inequality is not only taken into account but is also willingly reproduced in line with biological attributions concerning women and men.

10.3.2 Gendering of rationality/emotionality in the brain

I scrutinize the gendered connotations to the categories 'rationality' and 'emotionality' in further areas of the neurocultural discourse with a particular focus on the allocation of these categories to distinct brain areas. Against a long history of masculinized connotations to rationality in contrast to feminized connotations to emotionality (cf. Hagner 2008), recently, these attributions are again manifested with brain based arguments, as Simon Baron-Cohen's conception of the male S-ystemizer brain versus the female E-motional brain (Baron-Cohen 2003). The critical reflection of this concept and its distortions have recently been shown impressively by Nicole Karafyllis (2008), by Rebecca Jordan-Young (2010, 198ff), and by Cordelia Fine (2010, 199ff.). Here, I examine the gendered ideologies that are again intertwined with the field of social neurosciences when male rationality is conceived as being separated from female emotionality/empathy.

Tanja Singer and colleagues worked with some sort of Ultimate Game in which the participants of the study had to rely on counter partners to give back (or not to give back) some amount of money they had invested beforehand (Singer et al. 2006). The important part of the experiment, however, was the examination of fMRI scans that were taken after the game while the participants (16 women and 16 men) watched a previously fair or unfair counter player receiving a painful electroshock. Both gender groups showed activity in those areas that are classified as processing empathic reactions (Insula and Anterior Cingulate Cortex) during the punishment of fair counter players. But, as Singer pointed out, only females also mirrored empathic 'activation' during punishment of unfair players.

First, the knowledge production via neuroimaging procedure has to be questioned here. The allocation of 'empathic' activity in all conditions to females is a conclusion after processing a contrast analysis of activity between two task conditions (pain-no pain) in each particular setting (fair or unfair counter player). This data procedure results in a data reduction as it subtracts all activities that are activated in both conditions. Only the unique activity for the task condition remains visible in the image. This procedure is problematic insofar as it also masks all those activities that may be necessary under both conditions (cf. Roland & Zilles 1998, who called for intersection analyses instead of subtraction analyses). Subtraction analysis also may neglect seemingly 'noise' (however important this may be) compared to seemingly task-relevant activity (Haller & Bartsch 2009). Consequently, the visualization in the brain image only shows the additive activity under the pain-condition. In such, the brain image itself cannot solve the question, how much activity the participants – i.e., men included – produced under both conditions. Neither can the question be solved, whether men showed no empathic activation at all during punishment of unfair players, or how much of this activity had been subtracted by contrast analysis.

This study stated, in contrast, that male participants showed a higher 'feeling of revenge', assessed in a post-questionnaire. The questionnaire for stating these feelings of revenge ranged from –2 (not at all) to +2 (very much). The paper presents a graph with the y-axis scaling between 0 and +2, on which females on average reached a statement of revenge of 0.5 and men on average scaled 1.5 (Singer et al. 2006, 469). Notwithstanding that men stated higher feelings of revenge (on average 3.5 above 'not at all') females also did so, but to a lesser extent (on average 2.5 above 'not at all').[5] The presentation, however, exaggerated the impression of difference by showing only the upper part of the rating. There are analogous distortions in presentations of gender differences, e.g., in showing only the upper 20% of a 100% scale to underline differences and masking the amount of similarities between groups under analysis, an approach that was already criticized by Janet Hyde (Hyde 2005).

Additionally, an increased activation in men in the control area of reward (i.e., the Nucleus Accumbens, a region that is associated with the processing of reward) during punishment of unfair players was linked to their revenge preferences. As women also assessed feelings of revenge, but no activity was measured in their Nucleus Accumbens, this acknowledgement of revenge to this brain area turns out to be a bit difficult.

With a step further Singer's and her colleagues' interpretations on the results of their game study are worth for mentioning. Fairness and cooperative behavior in social interactions strengthen empathic binding, they conclude, but men rather tend to cut social relations to unfair partners whereas women, in general, react more emphatically and in an altruistic manner, irrelevant of the fairness or unfairness of others. There are some problematic aspects concerning these statements: (a) the generalization of the findings from Ultimate Games to explain complex empathic behavior, situated in 'real' world contexts, (b) the homogenization within the gender groups that cannot be assessed without information on the variation indices of the results, (c) the impacts that are drawn from these results for legitimizing gender roles and positions in society: "these findings could indicate a predominant role for males in the maintenance of justice and punishment of norm violation in human societies." (Singer et al. 2006, 468).

But men are not only framed as defenders of rational order, they are also set up as being hormonally driven (by their testosterone) and thus being more susceptible for financial risk behavior than females. Catherine Vidal has impressively worked out the constructions of linkages between sexed brains, testosterone levels, and economic return on the London trading floor in a recent study (Coates & Herbert 2008), and she uncovered the methodological failures and interpretative distortions of this paper, which was recently addressed several times with the slogan: "If Lehman Brothers had been Lehman Sisters, their financial performance would have been better" (Vidal 2012, 299).

10.4   Neurofeminist prospects for neurocultures

The analysis of the gendered cerebral subject has shown how the argumentative figures of individualization and self-responsibility in modulating the own materiality of the brain resource goes hand in hand with the encapsulation of gender differences within this brain's materiality; the latter hardly seems to be seen capable of such changes. My hypothesis is that the *disregard* of societal norms, structures, and discourses that cause gender hierarchies in a performative manner provides the persistent ascriptions of rationality and emotionality to the gendered subject. Or take it the other way round, the maintenance of a seemingly masculinized rationality and a feminized emotionality are to be taken as crucial to stabilize current societal order. Despite the emerging discourse of emotional capital on the labor market for women and men, the gendered connotations persist, i.e. men could and should uptake feminized emotional qualities and women could and should use masculinized rational skills (cf. Illouz 2008).

The permanency of sex-based ascriptions to rationality and emotionality is served to a great extent by the reference to their neuronal brain basics. How can a neurofeminist perspective elaborate a more differentiated research and discourse in these fields? Isabelle Dussauge and Anelis Kaiser have set up a list of demands for neuroscientific research to break up neuro-sexism (Dussauge & Kaiser 2012): to set up non-determinist research agendas, to conceptualize gender and brain development as a performative process within societal normative orders, to account for a variability of differences, and to reflect the aim of interpretations of research results. This is a first step to improve critical neuroscience.

Another important issue to question the notions of the gendered cerebral subject provided by contemporary neurocultures has to be realized in terms of the potentials for disrupting nature-culture dichotomies on both material and epistemological levels (Kraus 2012). Due to the strong entanglement between the brain's materiality and neurotechnological applications for reading, manipulating, and optimizing one's own biology, the neurocultural developments and concepts can also be read against the backdrop of the epistemological framework of feminist materialism that attempts to grasp materiality beyond essentialist determinism. Elisabeth Wilson, for example, elaborated a feminist materialism approach to account for the organic processes that are entangled with the use of anti-depressives. She speaks for addressing the biological processes in a more holistic view that includes body-brain intra-actions as well as therapeutic settings and normative social values concerning the framing of health and disease (Wilson 2008).

Cecilia Åsberg and Jennifer Lum have articulated a feminist materialist approach to cover the mutual entanglements between neuroscientific research and the gendered framing of Alzheimer's disease (Åsberg & Lum 2010). I analyze the intra-active communications in brain-computer interfaces by taking biological dynamics, brain plasticity, technical agents and codes, the role of actors and power relations, and the impacts of these agential realizations on the socio-cultural level into account (Schmitz forthcoming). These neurofeminist approaches will be continued to gain sound knowledge of the existing scientific controversies about sex/gender-related brain research and gendered neurocultures and to anticipate discrimination along the axes of gender and intersected categories.

Notes

1. According to gender research, sex is not a pure bodily or material fact, but is deeply interwoven with social and cultural constructions (cf. Fausto-Sterling 2012). Following this concept, the term sex/gender is used throughout this paper deliberately to emphasize the inextricable entanglements in such a bio-cultural approach.
2. Neuroethical approaches, i.e., a new interdisciplinary field of research that is settled within neurosciences, medicine and in ethical domains (with its own journals, e.g., *Neuroethics*), have outlined the pressure to conform in this framework and to criticize its harmful practices against those who are excluded with respect to equal access and distributive justice.
3. Ritalin with the active agent Methylphenidat (an amphetamine) is permitted for the treatment of ADHD and narcolepsy, and is also used in non-medical spheres for the improvement of attention and concentration. Prozac with the active agent Fluoxetin (blockage of uptake of serotonin) is permitted for treatment of depression and compulsory disorders; concerning bulimia, it is said to enhance positive mood.
4. This disregard also can be witnessed according to the side effects of Ritalin that include insomnia, increased irritability, restlessness, aggressive behavior, loss of appetite, weight loss, stomach problems, headaches, dizziness, dry mouth, and delayed growth in the long-term treatment of children. For Prozac side effects as nausea, weakness, indigestion, anxiety, nervousness, abdominal pain, weight gain, drowsiness, palpitations, thought disorder, tremor, sweating, headache, dizziness, dry mouth, and sexual dysfunction are listed.
5. There are no standard deviations given in the paper that would allow assessing the variability within the gender groups.

References

Alaimo, S. & Hekman, S. (2008). Introduction: Emerging Models of Materiality in Feminist Theory. In S. Alaimo & S. Hekman (eds), *Material Feminisms* (pp. 1–10). Bloomington: Indiana University Press.

Angel B., Pittenger, C., & Kandel, E.R. (2003). CREB, Memory Enhancement and the Treatment of Memory Disorders: Promises, Pitfalls and Prospects. *Informa Healthcare*, 7(1), 101–114.

Åsberg, C. & Lum, J. (2010). Picturizing the Scattered Ontologies of Alzheimer's Disease: Towards a Materialist Feminist Approach to Visual Technoscience Studies. *European Journal of Women's Studies*, 17(4), 323–345.

Barad, K. (2003). Posthumanist Performativity: Toward an Understanding of How Matter Comes to Matter. *Signs: Journal of Women in Culture and Society*, 28(3), 801–831.

Barad, K. (2007). *Meeting the Universe Halfway*. London: Duke University Press.

Baron-Cohen, S. (2003). *The Essential Difference: The Truth About the Male and Female Brain*. New York: Basic Books.

Beaulieu A. (2001). Voxels in the Brain: Neuroscience, Informatics and Changing Notions of Objectivity. *Social Studies of Science*, 31, 635–680.

Beaulieu, A. (2002). Images Are Not the (Only) Truth: Brain Mapping, Visual Knowledge, and Iconoclasm. *Science, Technology & Human Values*, 27(1), 53–86.

Bloch, C., Kaiser, A., Kuenzli, E., Zappatore, D., Haller, S., Franceschini, R., Luedi, G., Radue, E.W., & Nitsch, C. (2009). The Age of Second Language Acquisition Determines the Variability in Activation Elicited by Narration in Three Languages in Broca's and Wernicke's Area. *Neuropsychologia*, 47(3), 625–633

Bluhm, R., Jacobson, A.J., & Maibom, H.L. (2012). *Neurofeminism: Issues at the Intersection of Feminist Theory and Cognitive Science*. London: Palgrave Macmillan.

Blum, L.M. & Stracuzzi, N.F. (2004). Gender in the Prozac Nation: Popular Discourse and Productive Femininity. *Gender and Society*, 18(3), 269–286.

Boltanski, L. & Chiapello, E. (2005). *The New Spirit of Capitalism*. London: Verso.

Butler, J. (1993). *Bodies that Matter: On the Discursive Limits of Sex*. London: Routledge.

Coates, J.M. & Herbert, J. (2008). Endogenous Steroids and Financial Risk Taking on a London Trading Floor. *Proceedings of the National Academy of Sciences of the United States of America*, 105, 6167–6172.

Dolphijn, R. & van der Tuin, I. (2012). *New Materialism: Interviews & Cartographies*. Ann Arbor: Open Humanities Press (MPublishing, University of Michigan Library).

Draganski, B., Gaser, C., Busch, V., Schuierer, G., Bogdahn, U., & May A. (2004). Neuroplasticity: Changes in Grey Matter Induced by Training. *Nature*, 427, 311–312

Dussauge, I. & Kaiser, A. (2012). Re-queering the Brain. In R. Bluhm et al. (eds), *Neurofeminism: Issues at the Intersection of Feminist Theory and Cognitive Science* (pp. 121–144). UK: Palgrave Macmillan.

Farah, M.J., Illes, J., Cook-Deegan, R., Gardner, H., Kandel, E., King, P., Parens, E., Sahakian, B., & Wolpe, P.R. (2004). Neurocognitive Enhancement: What Can We Do and What Should We Do. *Nature Reviews Neuroscience*, 5, 421–425.

Fausto-Sterling, A. (2012). *Sex/Gender: Biology in a Social World.* New York: Routledge.

Fausto-Sterling, A., Coll, C.G., & Lamarre M. (2012a). Sexing the Baby: Part 1 – What Do We Really Know about Sex Differentiation in the First Three Years of Life? *Social Science & Medicine,* 74, 1684–1692.

Fausto-Sterling, A., Coll, C.G. & Lamarre M. (2012b). Sexing the Baby: Part 2 Applying Dynamic Systems Theory to the Emergences of Sex-related Differences in Infants and Toddlers. *Social Science & Medicine,* 74, 1693–1702

Fine, C. (2010). *Delusions of Gender: How Our Minds, Society, and Neurosexism Create Difference.* New York: Norton.

Fine, C. (2013). Is There Neurosexism in Functional Neuroimaging Investigations on Sex Differences? *Neuroethics,* 6, 369–409.

Fitsch, H. (2012). (A)e(s)th(et)ics of Brain Imaging. Visibilities and Sayabilities in Functional Magnetic Resonance Imaging. *Neuroethics,* 5(3), 275–283.

Fukuyama, F. (2003). *Our Posthuman Future: Consequences of the Biotechnology Revolution* New York: Pikador.

Greely, H., Sahakian, B., Harris, J., Kessler, R.C., Gazzaniga, M., Campbell, P., & Farah, M.J. (2008). Towards Responsible Use of Cognitive-enhancing Drugs by the Healthy. *Nature,* 456, 702–705.

Hagner, M. (2006). *Der Geist bei der Arbeit. Historische Untersuchungen zur Hirnforschung.* Göttingen: Wallstein.

Hagner, M. (2008). Genius, Gender, and Elite in the History of the Neurosciences. In N.C. Karafyllis & G. Ulshöfer (eds), *Sexualized Brains: Scientific Modelling of Emotional Intelligence from a Cultural Perspective* (pp. 53–68). Cambridge (Mass.): MIT Press.

Hall, W. (2004). Feeling "Better than Well." *EMBO Reports,* 5(12), 1105–1109.

Haller, S. & Bartsch, A.J. (2009). Pitfalls in fMRI. *European Radiology,* 19, 2689–2706.

Harding, S. (1987). *Feminism and Methodology: Social Science Issues.* Bloomington: Indiana University Press.

Höppner, G. & Schmitz, S. (2014). Neuroenhancement and Success: A Gendered Re-reading of Popular Media. In S. Schmitz & G. Höppner (eds), *Gendered NeuroCultures.* Vienna: Zaglossus (in press).

Hubel, D.H. & Wiesel, T.N. (1970). The Period of Susceptibility to the Physiological Effects of Unilateral Eye Closure in Kittens. *The Journal of Physiology,* 206(2), 419–436.

Hyde, J. (2005). The Gender Similarities Hypothesis. *The American Psychologist,* 60, 581–592.

Illouz, E. (2008). Emotional Capital, Therapeutic Language, and the Habitus of "the New Man". In N.C. Karafyllis & G. Ulshöfer (eds), *Sexualized Brains: Scientific Modelling of Emotional Intelligence from a Cultural Perspective* (pp. 151–178). Cambridge (Mass.): MIT Press.

Jenkins, W.M. & Merzenich, M.M. (1987). Reorganization of Neocortical Representations After Brain Injury: A Neurophysiological Model of the Bases of Recovery from Stroke. *Progress in Brain Research,* 71, 249–266.

Jordan-Young, R. (2010). *Brain Storm: The Flaws in the Science of Sex Differences.* Cambridge, Mass.: Harvard University Press.

Jordan-Young, R.M. & Rumiati, R.I. (2012). Hardwired for Sexism? Approaches to Sex/Gender in Neuroscience, *Neuroethics,* 5(3), 305–315.

Kaas, J.H., Kubitzer, L.A., Chino, Y.M., Langston, A.L., Polley, E.H., & Blair, N. (1990). Reorganization of Retinotopic Cortical Maps in Adult Mammals After Lesions of the Retina. *Science*, 248, 229–231.

Kaiser, A., Haller, S., Schmitz, S., & Nitsch, C. (2009). On Sex/Gender Related Similarities and Differences in fMRI Language Research. *Brain Research Reviews*, 61, 49–59.

Kaiser, A., Kuenzli, E., Zappatore, D., & Nitsch, C. (2007). On Females' Lateral and Males' Bilateral Activation During Language Production: A fMRI Study. *International Journal of Psychophysiology*, 63(2), 192–198.

Kandel, E.R., Schwartz, J.H., & Jessell, T.M. (2000). *Principles of Neural Science* (4th edn). New York: McGraw-Hill.

Karafyllis, N.C. (2008). Oneself as Another? Autism and Emotional Intelligence as Pop Science, and the Establishment of "Essential" Differences. In N.C. Karafyllis & G. Ulshöfer (eds), *Sexualized Brains: Scientific Modelling of Emotional Intelligence from a Cultural Perspective* (pp. 237–315). Cambridge, Mass.: MIT Press.

Kramer, P.D. (1994). *Listening to Prozac: The Landmark Book about Antidepressants and the Remaking of the Self.* New York: Penguin.

Kraus, C. (2012). Linking Neuroscience, Medicine, Gender and Society Through Controversy and Conflict Analysis: A "Dissensus Framework" for Feminist/ Queer Brain Science Studies. In R. Bluhm et al. (eds). *Neurofeminism: Issues at the Intersection of Feminist Theory and Cognitive Science* (pp. 193–215). London: Palgrave Macmillan.

Kuria, E.N. (2012). The Challenge of Gender Research in Neuroscience. In F. van der Valk (ed.). *Neuroscience and Political Theory* (pp. 268–288). New York: Routledge.

Latour, B. & Woolgar, S. (1979). *Laboratory Life: The Construction of Scientific Facts.* Princeton: Princeton University Press.

Maguire, E.M., Gadian, D.G., Johnsrude, I.S., Good, C.D., Ashburner, J., Frackowiak, R., & Frith, C.D. (2000). Navigation-related Structural Change in the Hippocampi of Taxi Drivers. *Proceedings of the National Academy of Science*, 97(6), 1–6.

Nikoleyczik, K. (2012) Towards Diffractive Transdisciplinarity: Integrating Gender Knowledge into the Practice of Neuroscientific Research. *Neuroethics*, 5(3), 231–245.

Normann, C. & Berger, M. (2008). Neuroenhancement: Status Quo and Perspectives. *European Archives of Psychiatry and Clinical Neuroscience*, 258(5), 110–114.

Ortega, F. & Vidal, F. (2007). Mapping the Cerebral Subject in Contemporary Culture. *Reciis*, 1(2), 255–259.

Pickersgill, M. (2013). The Social Life of the Brain: Neuroscience in Society. *Current Sociology*, 61(3), 322–340.

Pitts-Taylor, V. (2010). The Plastic Brain: Neoliberalism and the Neuronal Self. *Health: Interdisciplinary Studies in Health, Illness and Medicine*, 14(6), 635–652.

Pitts-Taylor, V. (2011). The NeuroCultures Manifesto. *Social Text* online, http://www.socialtextjournal.org/periscope/. Accessed July 25, 2013.

Roland, P.E. & Zilles, K. (1998). Structural Divisions and Functional Fields in the Human Cerebral Cortex. *Brain Research Reviews*, 26, 87–105.

Rose, N. (2007). *The Politics of Life itself: Biomedicine, Power, Subjectivity in the Twenty-first Century.* Princeton: Princeton University Press.

Rose, N. & Novas, C. (2005). Biological Citizenship. In A. Ong & S.J. Collier (eds), *Global Assemblages: Technology, Politics, and Ethics as Anthropological Problems* (pp. 439–463). Oxford: Blackwell.

Roy, D. (2012). Neuroethics, Gender and the Response to Difference. *Neuroethics*, 5(3), 217–230.

Sahakian, B. & Morein-Zamir, S. (2007). Professor's Little Helpers. *Nature*, 450, 1157–1159.

Schaper-Rinkel, P. (2012). Das neurowissenschaftliche Selbst. (Re)Produktion von Geschlecht in der neurowissenschaftlichen Gouvernementalität. In E. Sänger & M. Rödel (eds), *Biopolitik und Geschlechterverhältnisse*, Münster: Westfälisches Dampfboot.

Schmitz, S. (2010). Sex, Gender, and the Brain – Biological Determinism versus Socio-cultural Constructivism. In I. Klinge & C. Wiesemann (eds), *Gender and Sex in Biomedicine: Theories, Methodologies, Results* (pp. 57–76). Göttingen: Universitätsverlag Göttingen.

Schmitz, S. (2012). The Neuro-technological Cerebral Subject: Persistence of Implicit and Explicit Gender Norms in a Network of Change. *Neuroethics*, 5(3), 261–274.

Schmitz, S. (forthcoming). The Communicative Phenomenon of Brain-Computer-Interfaces. In V. Pitts-Taylor (ed.), *Mattering: Feminism, Science and Materialism*. New York: NYU Press

Singer, T., Seymour, B., O'Doherty, J.P., Stephan, K.E., Dolan, R.J., & Frith, C.D. (2006). Empathic Neural Responses Are Modulated by the Perceived Fairness of Others. *Nature*, 439, 466–469.

Singh, I. (2002). Bad Boys, Good Mothers, and the "Miracle" of Ritalin. *Science in Context*, 15, 577–603.

Sommer, I.E., Alemann, A., Bouma, A., & Kahn, R.S. (2004). Do Women Really Have More Bilateral Language Representation than Men? A Meta-analysis of Functional Imaging Studies. *Brain*, 127, 1845–1852.

van der Tuin, I. (2011). New Feminist Materialisms – Review Essay. *Women's Studies International Forum*, 34(4), 271–277.

Vidal, C. (2012). The Sexed Brain: Between Science and Ideology, *Neuroethics*, 5(3), 295–303.

Vidal, F. (2009). Brainhood, Anthropological Figure of Modernity. *History of the Human Sciences*, 22(19), 5–36.

Vul, E., Harris, C., Winkielmann, P., & Pashler, H. (2009). Puzzlingly High Correlations in fMRI Studies of Emotion, Personality, and Social Cognition. *Perspectives on Psychological Science*, 4, 274–290.

Wallentin, M. (2009). Putative Sex Differences in Verbal Abilities and Language Cortex: A Critical Review. *Brain and Language*, 108, 175–183.

Wilson, E.A. (1998). N*eural Geographies: Feminism and the Microstructure of Cognition*. New York: Routledge.

Wilson, E.A. (2008). Organic Empathy: Feminism, Psychopharmaceuticals, and the Embodiment of Depression. In S. Alaimo & S. Hekman (eds), *Material Feminisms* (pp. 373–399). Bloomington: Indiana University Press.

11
Non-Reductive Integration in Social Cognitive Neuroscience: Multiple Systems Model and Situated Concepts

Luc Faucher

In a companion piece to this paper (Faucher 2012), I adopted John Bickle's "new wave meta-scientific" stance (i.e. a bottom-up philosophy of science that tries to capture the sense of some concepts – like the concept of reduction – as it emerges from the sciences, "independent of any pre-theoretic or 'philosophical' account of these concepts" (Bickle 2003, 31)) and argued that reductionism does not capture the relationships between domains or theories in cognitive neuroscience. More precisely, I argued against the notion that classical reductionism as well as "meta-scientific reductionism" should apply across the board in neuroscience, and defended the idea that relations between fields (and theories inside them) in some parts of social cognitive neurosciences are *integrative* but in a *non-reductive* way. In this paper, I will illustrate this idea with an example drawn from social cognitive neuroscience (SCN henceforth) of racial prejudice. As I will demonstrate, in this field the nature of relationships between the disciplines that form the core of SCN has evolved over the years, without giving way to reduction. More precisely, I will show how establishing links between social psychology and cognitive neuroscience has allowed the formulation of a new hypothesis about the nature of implicit representations. In light of this, I will also propose that SCN of racial prejudice should consider some theoretical resources from a related field – that of "embodied" or "situated" cognition. I argue that by doing so it would gain additional theoretical unity (a form of theoretical progress) as well as the possibility of making new predictions (a form of empirical progress). I will suggest that borrowing constructs from another field is yet another illustration of the kind of non-reductive relationships characteristic of the type of integration sought by SCN.

Here's how I will proceed in what follows: I will first give a brief description of SCN (Section 1), then describe some of the on-going work in a sub-field of SCN, which focuses on racial prejudice (Section 2). I'll move on to explaining why reduction does not capture what is occurring in that part of SCN (Section 3), and then turn to some recent work by social neuroscientist David Amodio and his team on racial prejudice and demonstrate how one of their proposals illustrates the kind of non-reductive and practical relationships characteristic of mature SCN (Section 4). Finally (Section 5), I'll propose a framework that assists in the further understanding of the characteristics of prejudiced cognition as described by Amodio and his team. I will argue that this suggestion further illustrates the kind of contribution expected from neighboring disciplines in some fields of SNC.

11.1   What is Social Cognitive Neuroscience?

The term "Social Cognitive Neuroscience" (SCN) was introduced at the beginning of this millennium (Lieberman 2000; Ochsner & Lieberman 2001) to describe an interdisciplinary field of research that "seeks to understand phenomena in terms of the interactions between 3 levels of analysis: the social level...the cognitive level...and the neural level" (Ochsner & Lieberman 2001, 717). SCN is best viewed as a sub-field of Social Neuroscience (SN), where SN is understood to refer to any research that links the biological and social levels of analysis (Cacioppo & Berntson 2005, xiii; Harmon-Jones & Winkielman 2007, 3–4; Ochsner 2007, 43). For instance, SN includes work that spans the social to the physiological levels like Meaney and colleagues' research on the impact of the quality of maternal nurturance in young rat pups on hypothalamic-pituitary-adrenal reactivity in adulthood (Hellstrom et al. 2012) or Insel & Fernald's (2004) on the role of oxytocin and vasopressin in the formation of social bonds and social memory. The phrase "SCN" was coined by researchers who did not identify with this type of research. As Ochsner puts it:

> [T]he term "SCN" appealed to researchers who (1) were interested in using cognitive neuroscience methods to study a wide array of socioe-motional phenomena, (2) wanted to use this combined methodology to elucidate the information processing level of analysis...(Ochsner 2007, 43)

Thus what distinguishes SCN from other parts of SN is its focus on the informational-processing level and its use of CN's methodology. In terms of research goals,

> [s]ocial cognitive neuroscience has many of the same goals as social psychology in general, but it brings a different set of tools to bear on those scientific goals. (Lieberman 2007)

What are these goals? Cacioppo and Berntson have defined social psychology as "the scientific study of social behavior, with *an emphasis on understanding the individual in a social context.* Accordingly, social psychologists study a diverse range of topics ranging from intrapersonal processes shaped by or in response to others, such as the self, attitudes, emotions, social identity, normative beliefs, social perception, social cognition, and interpersonal attraction; to interpersonal processes such as persuasion and social influence, verbal and non-verbal communication, interpersonal relationships, altruism, and aggression, to group processes such as social facilitation, cooperation and competition; ... " (my emphasis; 2006, 91). In the late 1970s, social cognitive psychology emerged as a subfield of social psychology, focusing on information-processing accounts of the phenomena to be explained. Social cognition research's aim was to uncover the set of mechanisms and the type of representations underlying social inference and behavior. With the growing success of cognitive neuroscience, it was only a matter of time before a discipline (or sub-discipline) like SCN would emerge. As mentioned earlier, SCN is essentially social psychology (or more accurately social *cognitive* psychology) using the toolbox of cognitive neurosciences, such as fMRI and event-related potentials). As in cognitive neuroscience, researchers in SCN are using data provided by these tools to draw inferences about the nature and number of psychological mechanisms underlying social phenomena and the types of representation they use. Indeed, since the turn of the 21st century, equipped with these new tools, social psychologists have produced several functional imaging studies investigating the neural basis of psychological capacities implicated in social cognition. They have been able to show, for instance, that social pain resulting from exclusion shares the same neural basis as physical pain (Eisenberg & Lieberman 2004) or that some classes of moral judgments (personal ones) involve an emotional component, while others (impersonal ones) do not (Green et al. 2001).

Finally, let me insist on one very important meta-theoretical principle of SCN – one that might be forgotten if we focus exclusively on the content and method of SCN (Ochsner 2007, 52). As hinted by Cacioppo and Berntson in the previous paragraph's quote, social psychology studies social behaviors with an emphasis on understanding the individuals in "context." The notion is that the context is thought to affect the construal of the stimulus which the individual responds to; the context might be *internal* (the individual's temperament or motivation) and/or *external* (the presence of others or of some features of the environment). So the meaning of a stimulus is not inherent to it, depending only on its intrinsic features, but rather it is constructed and depends on the changing characteristics of the subject and/or of the situation. I'll later return to this idea as it will play a crucial role in my argument against reductionism in Section 3 as well as in the evaluation of the new model of racial prejudice cognition I propose in Section 5.

For the moment, I would like to describe some trends in the recent history of social psychology of racial prejudice, namely the emergence of an interest in "implicit prejudice" or "implicit bias." In the first years of its existence, the "neuroscience" component of SCN of racial biases was fairly conservative in almost exclusively proposing brain-mapping studies of psychological processes identified by social psychology. In this context, most of the theoretical progress came from the "social psychology" component of SCN. In recent years, the role of neuroscience has changed and some of the most interesting theoretical proposals are now coming from neuroscience. I will describe some of these proposals and argue that the relationships between theories in the context of SCN are non-reductive. Further along in this paper, I'll attempt to qualify these relationships more accurately.

11.2   SCN of racial prejudice

Work on prejudice has not escaped the trend described in the previous section. For more than 50 years now (since Allport's original work in 1954), social psychologists have been trying to explain the origin of many forms of prejudice, including racial prejudice. Under the heading "prejudice," social psychologists have included three phenomena that are often – but not always – correlated: stereotypes, prejudice per se, and discrimination. Studies of prejudice typically examine one of these three phenomena: the "cognitive" aspect (stereotypes, for instance, "all members of group X are lazy" or "all members of group X are good dancers"), the emotional or attitudinal aspect (prejudice per se, for instance, "I hate (or like) people of group X") and the behavioral aspect (discrimination, for instance, the fact that one preferentially hires members of group X). Stereotypes are typically responsible for biases in information gathering or in memory that are typical of racial cognition while prejudice is considered to be responsible for the affective quality of the reaction to outgroup or ingroup members. Both stereotypes and prejudices are thought affect behavior, though as we will see later [Section 4], their effects might be different.

One question that has become of crucial interest for social psychologists in recent years is the question of "implicit prejudice."[1] One reason social psychologists have become interested in implicit prejudice is that many people nowadays consider the expression of certain prejudices, for instance against people of other races, to be politically incorrect. Therefore, investigation of "explicit prejudice" is open to the possibility of a subject lying or deceiving. It is also possible that people are unconsciously influenced by the prevalent culture and sincerely express their beliefs when they claim to be non-prejudiced, while in fact being prejudiced (and discriminating) when interacting with people of other races. Indeed, these two ways of conceptualizing implicit prejudice correspond to the two different roots of implicit cognition (Payne & Garwonski 2010, 2): the first is found in research about selective attention and emphasizes the differences between automatic and controlled cognition, without being too concerned with the question of the consciousness of processes (Devine 1989; Fazio et al. 1995); the second is found in research about implicit memory and is more concerned with the question of consciousness and unconsciousness of processes (Greenwald & Banaji 1995). These different conceptualizations of the implicit are still debated in literature and there remains no consensus in view.

To tap into implicit prejudices (inaccessible to the experimenter either because of social desirability concerns on the part of the subject, or because they are unconscious) social psychologists had to developed "indirect" methods[2] (or borrow them from other fields) like semantic priming, implicit association tests (IAT), affect misattribution procedure, etc. (for a review of these methods, see Nosek et al. 2011).[3] It has been shown that these implicit methods outperform explicit ones in predicting behaviors, choices, or judgments in socially sensitive domains.

In recent years, we have witnessed a sustained effort to build neural models of implicit prejudices (see for instance Amodio 2008; Ito 2010; Ito & Bartholow 2009; Kubota et al. 2012; Stanley et al. 2008). I will describe in more details some of this work in upcoming sections (see also my 2012), but I will provide a glimpse of some results that have been generated by this research by presenting three examples. Notice that I did not pick these examples randomly; each will play a role in some of the arguments I will present later on.

1) Previous studies have identified the amygdala as one of the major structures in charge of the early and automatic evaluation of stimuli (Ledoux 1996). It had been showed that damage to the amygdala leads to the disruption of indirect, physiological expressions of fearful memories while leaving their explicit expression intact. Inspired by these findings, Phelps and her colleagues (2000) have

> used fMRI to examine the relationship between activation of the amygdala and behavioral measures of race bias. During brain imaging, white American subjects were shown pictures of unfamiliar black and white faces. They were simply asked to indicate when a face was repeated. After imaging, subjects were given a standard explicit assessment of race attitudes (the Modern Racism Scale or MRS; McConahay 1986) and two indirect assessments of race bias: Those subjects who showed greater negative bias on the indirect measures of race evaluation (IAT and startle) also showed greater amygdala activation to the black faces than to the white faces. This correlation between the indirect measures of bias and amygdala was not observed when the amygdala response was compared to the MRS, the explicit test of race attitudes (Phelps & Thomas 2003, 752).

In more recent work, Ito & Barholow (2009) and Stanley et al. (2012) have proposed situating the amygdala in a larger circuit involving (yet not exclusively) the Anterior Cingulate Cortex (ACC) and the Dorsolateral Prefrontal Cortex (DLPFC). The latter two structures respectively are thought to be involved (1) in the detection of conflict between automatic race attitudes and explicit goals or beliefs about racial equality (ACC) and (2) in the top-down implementation of control (DLPFC).

Two things should be noted that emerged from recent research. The first has to do with the malleability of implicit prejudice:

> In the early days of research using indirect measures of race attitudes, especially those that relied on automatic responses, researchers assumed that the outcome of racially negative attitudes was fixed. ... *However, emerging research demonstrates that race-based preferences, even those that are automatic, seem to be malleable and dependent on both situational and dispositional factors.* ... The situation can ... increase or decrease internal goals for reducing race-based discrimination (my emphasis; Kubata et al. 2012, 945)

The second concerns the fact that parts of the systems mentioned earlier might not be involved in every instance of implicit prejudice; therefore, there might sometimes be a disconnect between aspects of implicit prejudice (stereotype, attitudes, and discrimination). As Kubata and colleagues observe:

> even though amygdala activation consistently correlates with implicit race preference measures, damage to the amygdala only impairs performance on physiological measures of implicit preference, leaving performance on the IAT intact. (Kubota et al. 2012, 942)

2) The second set of results I wish to mention concern semantic knowledge. Semantic knowledge refers to the knowledge we have about objects or living things. For instance, you have semantic knowledge about bicycles: you know that they typically have two wheels, that they have a saddle to sit on, and that they can be used to travel from a location to another. This knowledge is helpful in that when you learn that a particular object is a bicycle (that it belongs to the category "bicycle"), you gain access immediately to knowledge about it: for instance, you know that you should be able to use it for transportation. When speaking about the knowledge we have about other human beings, social psychologists have also posited that we use semantic knowledge about them. In our interactions with human beings, we constantly use categorical knowledge or "stereotypes." For instance, if you learn that a person in front of you is French, you will presume that he or she enjoys quality cheese or that he or she likes baguette and wine.

It has usually been thought that semantic knowledge about people (stereotypes) is subserved by the same mechanisms that underlie semantic knowledge about objects and living things. In a recent paper, Contreras et al. (2012) argue that if this were true, the same brain regions should be active when one uses semantic knowledge about objects and living things (basically inferior frontal gyrus and inferotemporal cortex) and when one use stereotypes. What they found suggests that this is not the case:

> *knowledge about the characteristics of social groups bears little resemblance to knowledge about other (non-social) categories.* When participants made semantic judgments about a variety of non-social objects, brain regions traditionally associated with general semantics were engaged, including left inferior frontal gyrus and inferotemporal cortex. [...] *Instead, stereotypes activated a network of brain regions that have been*

linked regularly to tasks that involve social cognition, including extensive areas of MPFC [Medial Prefrontal Cortex], posterior cingulate, bilateral temporoparietal junction and anterior temporal cortex. (my emphasis; Cotreras et al. 2012, 768)

What these researchers conclude from their results is the "category-specific" nature of social semantic knowledge and the fact that social semantic knowledge shares more with our representation of mental states than with semantic knowledge of non-social semantic categories.

3) In a recent series of papers, Meyer (Meyer et al. 2012) has suggested the existence of a social working memory different from (but complementary to) working memory. It is generally thought that working memory's function is to "build up and maintain an internal model of the immediate environment" (Bower, quoted by Meyer et al. 2012, 1887. It makes sense for researchers to postulate that we also need an internal model of our immediate "social environment" in order to navigate fluidly in it (such a model would help agents adopt attitudes and actions appropriate to the target). It is also believed that there is resource competition between working memory and the mentalizing system, such that when working memory load increases, the activity of the mentalizing system decreases.

What Meyer has found is that when one uses social stimuli in tasks, not only are the neural systems involved in working memory activated, but the mentalizing network is also activated. Moreover, the more demanding a task is, the more regions of the mentalizing network (the medial frontoparietal regions and the tempoparietal junction) are active – such as the "only mentalizing regions' parametric increases correlated with trait differences in perspective-taking ability" (2012, 1885). They thus conclude that there is a social working memory whose function is to build up and maintain an internal model of the immediate social environment.

*   *   *

As I stated earlier, this is just a glimpse of the kind of work underway in the SCN of racial prejudice. I could have presented additional other findings, but these examples should be sufficient to illustrate the point I wish to make in the next section, which is that in the SCN of racial prejudice, the relationships between different fields and theories in those fields are not reductive.

11.3   Reduction and non-reductive integration

In the companion piece to this paper (Faucher 2012), I argued against Bickle's idea that "ruthless reductionism,"[4] a form of *competitive pluralism*,[5] rules in neuroscience. As I then pointed out, SCN incarnate is a different form of pluralism, *compatible pluralism* (that is, the idea that a phenomenon in a domain can or must be explained by different theories – or theories at different levels, what I termed *"interlevel compatible pluralism"*). One could go further and say that SCN proposes a form of *"collaborative interlevel pluralism,"* where different theories or fields at different levels are not only compatible (in that they do not make propositions that are logically inconsistent with one another), but that they collaborate together to explain a phenomenon or a group of phenomena. One could see this attitude as built-into SCN since "[t]he goal of cognitive neuroscience research is to construct a theory of a functional architecture that describes a phenomenon at these three levels of analysis" (Ochsner 2007, 49). After all, the phrase "SCN" was introduced by people who did not identify with the type of research questions and methodology of those inhabiting more reductionist quarters of SN.

Since I wrote my initial paper, Bickle's position has changed (see his 2012), and now he is more than willing to admit that some parts of neuroscience are defined by compatible pluralism (or by collaborative interlevel pluralism). He now argues that neuroscience does not speak in one voice, and that different fields have different views about relationships between theories. This is precisely what I had tried to argue in my paper. While I am more than happy with his admission, I am not entirely satisfied. In that previous paper, I was also trying to argue that we should be more precise in our description of the nature of relationships between theories. So, even if Bickle and I now agree on the fact that there are some quarters of neuroscience that are non-reductionist, only part of the job is done – we still have to characterize these relationships. In order to do that, I borrowed another distinction, introduced by Todd Grantham (2004).

Grantham proposed that when discussing the integration of divergent theories in a single field (like SCN), we should consider the different reasons that motivate the integration (and therefore, different types of connection between theories). Sometimes the reasons are *theoretical*: one wants to *reduce* a higher-level phenomenon to a lower-level phenomenon; sometimes one might also want to *refine* theoretical constructs (this is what Contreras et al. 2012 and Meyer et al. 2012 are doing by distinguishing semantic knowledge from social semantic knowledge and working memory from social working memory, respectively) or *extend* the explanation (that is, provide an explanation of a phenomenon that theories at a higher level cannot provide). Other times, reasons are *practical*: for instance, one might establish connections between theories for *heuristic reasons* (in order to generate new hypotheses), for *confirmation reasons* (because work in lower-level fields would confirm theses in higher-level fields), or for *methodological purposes* (because methods from one field are more apt to perform a particular task than methods from the field that initially studied the phenomena). I argued that in the case of the SCN of prejudice, connections between theories are both theoretical (though not reductive) *and* practical.

One might mistakenly think that the SCN of prejudice (or SCN in general for that matter) is reductionistic because most papers in the field report brain mapping studies, that is, studies that ask "where in the brain is…[insert psychological construct here]?" (Amodio 2010, 698). Though, it is well acknowledged in the field that brain mapping is merely a first step in the study of the phenomena of interest for the SCN of prejudice. For instance, Amodio states that

> The hypothesis-testing approach in social neuroscience is used to test hypotheses about psychological variables. It begins with the assumption that a particular brain region reflects a specific psychological process. In this regard, *it does not concern brain mapping*, but instead relies on past brain-mapping studies to have already established the validity of neural indicators. (my emphasis; Amodio 2010, 699; see also Ochsner & Lieberman 2001, 729; Fiske 2000, 314)

The claim I was making in my (2012) paper was that the principal function of brain mapping in SCN is to contribute to building what I called a "heuristic model" (or a "neuro-functional model") of prejudiced cognition. Construction of this model requires the integration of knowledge gathered by social cognitive psychologists and cognitive neuroscience researchers (in their case, not necessarily knowledge concerning psychological capacities related to social phenomena, for instance, knowledge concerning non-social fear). Once such a model has been constructed, researchers can use it, for instance, to assess psychological variables that are not (or are not as easily) measured by other means (Harmon-Jones & Winkielman 2007, 6). They may also use this model to generate and test new theoretical hypotheses. Amodio points to such an idea when he writes:

> The greatest power of the hypothesis testing approach is that it establishes a channel of communication between social psychology and

other neuroscience-related fields...Through this link to the neuro-science literature on fear conditioning, researchers could begin to apply the vast body of knowledge on this type of learning and memory to form new hypotheses for how implicit affective racial biases are acquired, expressed in behavior, and potentially reduced. (2010, 700)

Amodio refers here to the work of Olsson and Phelps on "fear condi-tioning" of social targets (like members of particular social groups; see Olsson et al. 2005). In other words, he points to the fact that links between theories in SCN have a dynamic character: one starts with building a neuro-functional architecture (which requires brain-mapping), then once this architecture is provided, one can exploit it to generate and test new hypotheses. This shows clearly that one of the ultimate goals of SCN in terms of brain mapping is not reduction (by localization of psychological mechanisms), but rather the generation and testing of new hypotheses. In the next section (Section 4), I want to present another example of this "second wave" of work in the SCN of prejudice, that is, work that uses brain mapping to generate new hypotheses. Yet prior to this, I want to make a comment based on my earlier description of SCN.

In my companion paper, I argued that Bickle's ruthless reductionism was not descriptively adequate for some parts of the neurosciences. I want to insist here on a different reason why ruthless reductionism is not adequate. As we have seen earlier, one element of the meta-theory of SCN is the fact that social behavior is the result of the interaction between a person and a situation. As Ochsner notes: "Critically, it implies that the meaning we ascribe to a stimulus is not inherent in the stimulus itself but, rather, is a flexible product of our interpretation or *construal* of its meaning according to our current goals, which in turn may be a function of the current context" (2007, 49). Ochsner illustrates the lack of awareness of this metatheoretical element by considering studies of emotion. Many of these studies use facial expression or different sensory stimuli (auditory or olfactory, for instance). As meta-analysis reveals, these studies are receiving mixed results. According to Ochsner, the reason for this is that "these studies treated emotion as stimulus prop-erty like shape, size, or color rather than a context-dependent appraisal process that interprets the emotional value the stimulus in context of an individual's current goals, wants, and needs. As a consequence, these studies fail to manipulate the way in which individuals construed, or appraised, the meaning of stimuli, which leaves participants free to appraise stimuli in numerous ways" (idem, 50). Construal of stimuli is especially important for the SCN of prejudice. Study after study (for instance, Johnson & Fredrickson, 2005; Wheeler & Fiske, 2005), it has been shown that the way an individual construes another individual (as belonging to group A or as being a successful professional, or as a father of a big family) has an impact on the neural processes that will be activated. It has also been shown that the context (competition, failing a test, dirtiness, etc.) has an impact on how an individual will perceive others. Due to the central role of the construal and because of the nature of the factors (mainly, but not exclusively, social and psychological) that influence it, it is unlikely that an explanation of the phenomenon on which SCN focuses could be exclusively reductionistic.

As we have seen in the present section, the SCN of racial prejudice has both built-in goals and built-in assumptions that preclude it from adopting a reductive stance towards the psychological phenomena it studies. In the next section, I will present an example of the second wave of work in the SCN of racial prejudice, and I will propose a model of it that – if accepted – will provide ammunition to people who, like me, support a non-reductive yet integrative picture of the relationships between theories or fields in SCN.

11.4   Multiple memory systems as a model

In what looks now like a premonitory remark, Ochsner and Lieberman proposed that

> *research on stereotype representations could benefit from a research program similar to the one that has dissected memory representations.* Although the past decade has seen increasing acceptance of the notion of implicit stereotypes, little has been done to make finer representational distinctions at both the explicit and implicit levels. *Presumably, there are episodic, semantic, perceptual, affective and procedural stereotype representations.* Each kind of stereotype representation could have different constraints in terms of initial formation, activation, application, controllability, and capacity to be extinguished (my emphasis; 2001, 727; see also 700).

Though this idea found echoes in some researchers' reflections at the time, it was only systematically explored recently by Amodio and colleagues in a series of papers (Amodio 2008; Amodio & Devine 2006; Amodio, Harmon-Jones, & Devine 2003; Amodio & Mendoza 2010;

Amodio & Ratner 2011). In this section I'll present their proposal, and in the next, I'll propose a tentative framework which if adopted would contribute greatly to SCN's progress (both theoretical and empirical).

In one of the first of this series of papers, Amodio and Devine (2006) argued that the SCN of prejudice has failed to distinguish the cognitive and affective components of prejudice, and that, for this reason, the interpretation of their results remains elusive.[6] In their view, one must distinguish at least two types of implicit biases, namely *implicit evaluations* and *implicit stereotypes*. Implicit evaluations are negative affective responses directed at members of specific racial groups; while implicit stereotypes are culturally-shared beliefs about members of these racial groups. Amodio and Devine demonstrated that implicit evaluations and implicit stereotypes have different effects. These researchers posited that implicit evaluations are good predictors of negative behaviors. For example, they predict the frequency of unfriendly attitudes towards members of other races; they also predict the frequency of uncomfortable interactions, which may manifest themselves in hesitations, increased blinking in conversations with individuals belonging to another race, or a tendency to sit further away from the members of other races. On the other hand, implicit stereotypes predict cognitive effects, such as memory biases (for example, remembering more easily that a violent event was perpetrated by a Black individual than by a White individual) or perceptive biases (for example, finding the same behavior more violent when perpetrated by a Black versus a White individual).

The distinction established between affective and semantic forms of learning and memory in the neuroscientific literature gives ground to Amodio and Devine to postulate that implicit evaluation and implicit stereotyping might reflect the workings of two different systems. In their view, evaluations and stereotypes would depend on different cerebral structures: the amygdala for the former and more recent evolutionary cortical zones for stereotypes.

More recently, Amodio and Ratner (2011) have proposed what they have called the "Memory Systems Model" (MSM) of implicit racial cognition. This model rejects as well single-system models of implicit cognition (i.e. models that posit implicit processes "reflect a single system of symbolic or connectionist representations of semantic information in memory" (143)). This time, taking their cue from the cognitive neuroscience of memory,[7] which has distinguished several forms of non-declarative memory[8] and identified different locations in the brain as their substrates, they propose that implicit social cognition should be thought of along the same lines. In their paper, they propose considering a subset of the systems that could underlie implicit social cognition. These systems are the semantic associative memory (in charge of forming associations between cognitive concepts, for instance between a target group and a property – Asians and mathematics), fear memory (in charge of forming associations between a target group and an affective response – White and fear) and instrumental learning and memory (in charge of forming associations between a target group and an action tendency – Black and avoidance). It is postulated that each system has different properties (that is, racial biases are learned, stored, and expressed differently by each system) and has different neural substrates (neocortex – semantic memory; amygdala – fear memory; and striatum and basal ganglia – instrumental memory). Research has already begun to confirm the MSM (Olsson et al. 2005) and has led to important insights concerning the acquisition and the potential ways of getting rid of some of types of associations (Shurick et al. 2012). The adoption of MSM might lead to further discoveries[9] and distinctions, as the study of implicit racial prejudice begins to be tied more intimately to the CN of memory. Progress in the latter field might lead to progress in the previous field (and vice versa).

One final note to conclude this section. While Amodio and his colleagues have emphasized distinctions between different forms of implicit memory systems, it is believed that these systems typically work in concert to produce social behavior.[10] Though they are not keen on furnishing a model of the interaction between these systems, such a model should be provided to understand social cognition "in the wild." This is precisely such a model that what I want to propose in the next section.

11.5   Socially situated and embedded theory of concepts

The model of implicit racial prejudice I will propose is a version of the "socially situated and embedded theory of concepts." As Semin and Smith (2002) observed, despite having precedents in the field (see Carlston 1994), it has been championed by a few isolated voices and has never been really integrated to the field of social psychology. I want to argue that a better integration of this model would not only help to make sense of current findings (which can be considered as a form of *theoretical progress*), but would also contribute to SCN's *empirical progress* by helping to formulate novel predictions. In introducing this model, I will be following Niedenthal and his colleagues' (Niedenthal et al. 2005;

but also Barsalou 2008) description of the Perceptual Symbol System (PSS) theory.

This socially situated and embedded theory of concepts is better understood in contrast with the usual view of concepts. According to the classic view on concepts, concepts are acquired by abstraction from percepts. In the process of abstraction, these mental representations become stripped away from traces of the different sense modalities they are abstracted from (this is, among other things, what is thought to distinguish episodic memory from semantic memory). In other words, they are *amodal*. It is also postulated that these abstract representations are *modular* (they are autonomous from episodic memory as well as other systems, like affective memory), *decontextualized* (the process of categorization abstracts away from situations to focus on a common core), and *stable* (the same knowledge is used in all and every situation) (Barsalou 2008, 237). In contrast, a situated and embodied theory of concepts asserts that concepts use "partial reactivations of states in sensory, motor, and affective systems to do their jobs... The brain captures modality-specific states during perception, action and interoception and then reinstates parts of the same states to represent knowledge when needed" (Niedenthal, 2007, 1003). Representations are then modal (they are realized by modality-related re-activations), not modular (in that the conceptual system is not autonomous from other systems), contextualized (in that conceptualizations are linked to situations or contexts), and fluid (in that concepts deliver different packages of information in different contexts or situations).

Among the central constructs of PSS are the concepts of "simulator," "simulations," and "situated concepts." I want to introduce these concepts through two examples. Let's take first the concept of "rage." According to PSS (see Niedenthal 2007 for instance), it is represented in a multimodal fashion, including sensory, motivational, motor, and soma-to-sensory features which are typical of episodes of rage. It also contains or it is linked to information about situations in which rage is typically produced.[11] Lindquist (forthcoming) describes one way these representations are tied together (there might be other ways that do not involve language): "Emotion categories might be acquired in childhood by boot-strapping situations and core affective feeling to the words used by adult caregivers (e.g., when Mom and Dad tell Joey not to be 'sad', because of a broken toy, Joey learns that negative feelings following a loss are associated with a category 'sadness' in his culture)" (Lindquist, forthcoming).

All these associated representations form a 'simulator' which can create 'on the fly' representations adapted to particular instances of a category ('simulations').[12] For example, in some contexts, our mental representation of "rage" might include a strong visceral reaction, but this might not be included in others. Thus, "situated conceptualizations" of rage will include only properties that are contextually relevant. For instance, when thinking about "rage" in the context of "road rage," one will represent a loss of control, swearing, violent actions, etc.; while thinking about the "rage" of an athlete who wants win an event, one's depiction will include looks of determination on their face, concentration, stress, etc. Conceptualizations are thus *situated* in that a simulator will deliver a different package of information (a simulation) in different situations. As Barsalou puts it: "From this perspective [the perspective of PSS], a concept is neither a static database nor a single abstraction. Instead, it is an ability or competence to produce specialized category representations that support goal pursuit in the current setting, where each specialized representation is akin to an instruction manual for interacting with a particular category member" (Barsalou 2008, 144). Over time, the situated conceptualization comes to mind automatically when a particular situation is detected. One might think that these automatic situated conceptualizations are exactly what social psychologists are referring to when they talk about "implicit cognition."

The second example is borrowed from Neidenthal et al. (2005). Taking the concept of "politician," they say, "[f]ollowing exposure to different politicians, visual information about how typical politicians look (i.e., based on their typical age, sex, and role constraints on their dress and their facial expressions) becomes integrated in a simulator, along with auditory information for how they typically sound when they talk (or scream or grovel), motor programs for interacting with them, typical emotion responses induced in interactions or exposures to them, and so forth" (Niedenthal et al. 2005, 195). The concept (simulator) of politician will deliver different information in different contexts (situated simulation): in an electoral campaign, a politician may be more likely to appear all smiles, in a shirt with rolled up sleeves and a shovel in their hands, making promises best not believed; but while delivering a financial budget, they will appear to be deadly serious, wearing a coat, and possibly new shoes (in certain parliamentary traditions).

Let's return now to implicit prejudice. As was seen in Section 4, different memory systems are involved in stocking information about a category: semantic, affective, and instrumental. PSS invites us to think that in real or "wild cognition" these different systems are integrated. As I mentioned earlier, this suggestion has precedents: for instance,

Carlston's (1994) Associated Systems Theory (AST). According to his theory:

> Person impressions as well as attitudes are [...] [representations that] include several different types of information that reflect contributions from four underlying representational systems: visual, verbal, affective and action. For example, an impression may include images of the person's appearance (visual system), emotions felt toward the person (affective system), and traits believed to characterize the person. Importantly, the impression also includes representations of the perceiver's own behavior toward the person (action system), such as giving the person hugs or teasing him. (Semin & Smith 2002, 387).

PSS shares most of AST's assumptions, except the fact that the latter gives primacy to verbal information over other forms of information, a primacy which PSS rejects (but we aren't concerned with this aspect of the theory here).

Thus for PSS, our concept (simulator) of a social group (a race or a gender) will contain information not only about characteristic traits of the group ('Xs are good dancers but bad curlers'), but also auditory and visual representations as well as characteristic affective reactions and motor reactions towards member of it. This way of thinking about concept has the advantage of explaining the link between the concept and action in a way that is not otherwise explained, or not explained at all, by an abstract, amodal view of stereotypical representations. For instance, studies (for example, Chen & Bargh 1999) have documented that if one is required to pull a lever towards or away from oneself to report the valence of a word, the movement used will affect the speed of response: the response is faster for a positive word if the lever motion is towards oneself, and slower the lever is pushed away (and vice versa for negative words). This is supposed to show that there are associations between "self" and "positive" and "other" and "negative." But it also shows that this association is not (only) abstract, it also comprises a motor element (and is probably linked with instrumental memory). It also explains why damage to the amygdala does not affect IAT, but only physiological responses (see Section 2 example 1). In this case, the affective part of the concept is missing, but other aspects are present – and it would be interesting to know which ones. The reason this question is important is because PSS distinguishes between different ways of representing concepts. This is made clear by a distinction introduced by Niedenthal and colleagues (2005, 199) between *shallow* and *deep* processing. According to them, a task needs deep processing when it requires a simulation for its completion. When it does not, that is, when it can perform a task by only using word-level representation, it is said to be shallow. So one important question when one performs an IAT, for instance, is: what am I tapping into? Is this association I am detecting the reflection of a "superficial," purely linguistic, representation or is it the reflection of something deeper – a partial simulation – that could have direct impact on future action? For instance, it has been shown in consumer attitudes and behavior research that

> measure tapping implicit emotions evoked by consumer products predicted participants' buying intentions significantly better than their general implicit attitudes toward those products. Interestingly, this finding nicely parallel intergroup emotions literature which also makes the case that the specific emotions people feel toward various outgroups predict their action tendencies better than global evaluations, probably because emotions are more tightly linked to specific goal-directed action tendencies compared to global evaluations. (Dasgupta 2010, 55)

Because affective elements are only one part of the conceptual compound, it would also be important to develop tests that could discriminate the other elements (motor tendencies, etc.). PSS also clears the way for experiments where emotions or motor reactions are not allowed to be expressed. Could this have an impact on the categorizations tasks? On behavior towards outgroups? Moreover is the superficial/deep distinction following the lines of semantic knowledge/social semantic knowledge postulated by Contreras and colleagues such that if one thinks superficially about a member of a category, mentalizing systems are less likely to kick in?

Finally, PSS conceptualization fits well with a theme of social cognition that is gaining momentum nowadays: the malleability of implicit attitudes.[13] This malleability could be explained by the fact that "[h]uman cognitive systems produce situated versions of concepts that have context-specific functions rather than activating the same, context-independent configuration in every situation" (Smith & Semin 2007, 134). As was seen earlier, PSS predicts that simulators will deliver different packages of information depending on context. For instance, if an individual is angry or afraid for irrelevant reasons, negative stereotypes will be more readily applied to out-group members than when that individual is in a neutral or happy state (ibid., 133). Similarly, in a dark alley or when cues of vulnerability to disease are made salient, specific prejudices concerning some group are activated, while others, applicable to the same group, are not. For instance, as Schaller and Conway (2004) report that "[d]arkness amplified prejudicial beliefs about danger-relevant traits (trustworthiness and hostility), but did not much affect beliefs about equally-derogatory traits less relevant to danger" (155). Dasgupta and colleagues (2009) report similar findings, showing that "[r]ather than serving as a general warning that orient perceivers to generic dangers thereby increasing bias against any outgroup, emotions exacerbate bias only when the feeling warns of a specific threat that is directly applicable to the outgroup being appraised" (589).

These are only few illustrations of the potential of progress that would follow the adoption of an embedded and situated theory of concepts in SCN of racial prejudice. I contend that the adoption of this model could lead to major breakthroughs in this field.

11.6   Conclusion

In this paper, I adopted Bickle's meta-scientific stance and argued in favor of a non-reductionistic picture of relationships between theories and fields in some quarters of neuroscience. Contrary to what Bickle has once suggested, there is no (descriptive or normative) reason to reduce relationships between theories or fields of different levels to (ruthless) reduction. I also argued that for reasons linked with the meta-theory of SCN, more precisely because of the influence of context on cognition, reduction was quite unlikely. I also gave an illustration of the nature of relationships between the disciplines that form the core of SCN has evolved over the years. I have showed that this dynamic did not give way to reduction; it did not diminish the explanatory importance of social or cognitive psychology, but fueled these domains with exciting new ideas. More precisely, I have showed how in establishing links between social psychology and cognitive neuroscience of memory, some researchers in SCN have allowed the generation of a new hypothesis about the nature of implicit representations. I also proposed that SCN of racial prejudice should consider the theoretical resources offered by the related field of "embodied" or "situated" cognition. I argued that the adoption of such a model would lead to progress in SCN. I have suggested that this is also an illustration of the non-reductive relationships that are to be expected between different fields in CN and SCN.

Notes

1. As Payne and Gawronski put it: "Within the space of two decades, virtually *every intellectual question in social psychology*, and many outside of it, has been shaped by theories and methods of implicit social cognition" (my emphasis; 2010, 1).

2. The term 'indirect' is used to refer to the features of measurement procedures that "provide indicators of psychological attributes (e.g., attitudes) without having to ask participants to verbally report the desired information" (Payne & Gawronski 2010, 4).

3. The interest in developing windows into social psychological processes is also foundational in social neuroscience and is one of the historical roots of SCN of prejudice (see Harmon-Jones & Winkielman 2007, 4).

4. "Ruthless reductionism" describes and makes explicit the trend Bickle observes in neuroscience where theories at higher levels of explanation are replaced, without loss, by theories at lower levels such that, for instance, psychological phenomena can be explained directly at the molecular level. See the discussion in Faucher and Poirier 2013.

5. The expression "competitive" and "compatible" pluralism are borrowed from Mitchell (2002).

6. For a similar complaint about some earlier studies, see Judd et al. 2004, 76.

7. For an overview of this type of work see Squire & Wixted 2011.

8. Declarative memory refers to consciously accessible memory, while non-declarative memory refers to memory that is not consciously accessible, but that could be expressed in behavioral tasks.

9. For instance, the explanation of individual differences in the acquisition of some types of racial biases, see Levingston & Drwecki 2007.

10. As they themselves admit: "Implicit attitudes accessed by sequential priming tasks may reflect a combination of semantic associations (e.g., with good vs. Bad concepts), threat- or reward-related affective associations, and instrumental associations (e.g., reinforced and habitual actions)" (Amodio & Ratner 2011, 145).

11. Barsalou insists that for PSS the simulator for a category does not comprise background situations, but is linked to simulators of situations or events. According to Barsalou, a particular simulation is the result of the interaction between the simulator of a particular category of objects or people, for instance "politician," and a simulator for a category of events or situations, for instance "electoral campaign" or "budget announcement."

12. As Niedenthal and colleagues put it: "According to PSS, the simulation process is highly dynamic and context dependent. ... Depending on the current state of the simulator, the current state of associated simulators, the current state of broader cognitive processing, and so forth, a unique simulation results" (2005, 196).

13. "In the eyes of construction theorists, the high malleability of indirect measurement scores confirmed their assumption that contexts influence what information is used to construct an attitude from one moment to the next, and that these principles apply equally to direct and indirect measurement procedures. In fact, the very idea that indirect measurement procedures would assess rigid 'things' in memory independent of the context was seen as ill founded" (Payne & Garwonski 2010, 6).

References

Allport, G. (1954). *The Nature of Prejudice*. Reading, MA: Addison-Wesley Publishing Co.

Amodio, D.M. (2008). The Social Neuroscience of Intergroup Relations. *European Review of Social Psychology*, 19, 1–54.

Amodio, D.M. (2010). Can Neuroscience Advance Social Psychological Theory? Social Neuroscience for the Behavioral Social Psychologist. *Social Cognition*, 28(6), 695–716.

Amodio, D.M. & P.G. Devine. (2006). Stereotyping and Evaluation in Implicit Race Bias: Evidence for Independent Constructs and Unique Effects on Behavior. *Journal of Personality and Social Psychology*, 91(4), 652–661.

Amodio, D.M., E. Harmon-Jones, & P.G. Devine. (2003). Individual Differences in the Activation and Control of Affective Race Bias as Assessed by Startle Eyeblink Responses and Self-Report. *Journal of Personality and Social Psychology*, 84, 738–753.

Amodio, D.M. & S.A. Mendoza. (2010). Implicit Intergroup Bias: Cognitive, Affective, and Motivational Underpinnings. In B. Gawronski & K. Payne (eds), *Handbook of Implicit Social Cognition* (pp. 353–374), New York: Guilford.

Amodio, D.M. & K.G. Ratner. (2011). A Memory Systems Model of Implicit Social Cognition. *Current Directions in Psychological Science*, 20(3), 143–148.

Barsalou, L. (2008). Situating Concepts. In P. Robbins and M. Ayede (eds), *Cambridge Handbook of Situated Cognition* (pp. 236–263). New York: Cambridge University Press.

Bickle, J. (2003). *Philosophy and Neuroscience: A Ruthlessly Reductive Account*. Dordrecht: Kluwer Academic Publishers.

Bickle, J. (2012). A Brief History of Neurosciences's Actual Influences on Mind-Brain Reductionism. In S. Gozzano & C. Hill (eds), *New Perspectives on Type Identity Theory* (pp. 88–109). Cambridge: Cambridge University Press, 2012.

Cacioppo, J.T. & G.G. Berntson. (2005). Analyses of the Social Brain through the Lens of Human Brain Imaging. In J.T. Cacioppo & G.G. Berntson (eds), *Social Neuroscience* (pp. 1–17), New York: Psychology Press.

Cacioppo, J.T. & G.G. Berntson. (2006). A Bridge Linking Social Psychology and the Neuroscience. In P.A.M. Van Lange (Ed.), *Bridging Social Psychology: The Benefits of Transdisciplinary Approaches* (pp. 92–96), Hillsdale, NJ: Erlbaum.

Carlston, D. (1994). Associated Systems Theory: A Systematic Approach to Cognitive Representations of Persons. In R.S. Wyer and T.K. Srull (eds), *Advances in Social Cognition*, vol. 7, pp. 1–78. Hillsdale, NJ: Lawrence Erlbaum Associates.

Chen, M. & J. Bargh. (1999). Nonconscious Approach and Avoidance Behavioral Consequences of the Automatic Evaluation Effect. *Personality and Social Psychology Bulletin*, 25, 215–224.

Contreras, J.M., M.R. Banaji, & J.P. Mitchell. (2012). Dissociable Neural Correlates of Stereotypes and Other Forms of Semantic Knowledge. *SCAN*, 7, 764–770.

Dasgupta, N. (2010). Implicit Measures of Social Cognition: Common Themes and Unresolved Questions. *Journal of Psychology*, 218(1), 54–57.

Dasgupta, N., D. DeSteno, L.A. Williams, & M. Hunsinger. (2009). Fanning the Flames of Prejudice: The Influence of Specific Incidental Emotions on Implicit Prejudice. *Emotion*, 9(4), 585–591.

Devine, P. (1989). Stereotypes and Prejudice: Their Automatic and Controlled Components. *Journal of Personality and Social Psychology*, 56(1), 5–18.

Eisenberg, N.I. & M.D. Lieberman. (2004). Why Rejection Hurts: A Common Neural Alarm System for Physical and Social Pain. *Trends in Cognitive Sciences*, 8(7), 294–300.

Faucher, L. (2012). Unity of Science and Pluralism: Cognitive Neuroscience of Racial Prejudice as a Case Study. In O. Pombo, J.M. Torres, J. Symons, & S. Rahman (eds), *Special Sciences and the Unity of Science* (pp. 177–204), Dordrecht: Springer.

Faucher, L. & P. Poirier. 2013. Le nouveau réductionnisme "nouvelle vague" de John Bickle. In M. Silberstein (Ed.), *Matériaux Philosophiques et Scientifiques pour un Matérialisme Contemporain* (271–317), Paris: Éditions Matériologiques.

Fiske, S. (2000). Stereotyping, Prejudice, and Discrimination at the Seam between the Centuries: Evolution, Culture, Mind, and Brain. *European Journal of Social Psychology*, 30, 299–322.

Fazio, R.H., J.R. Jackson, B.C. Dunton, & C.J. Williams. (1995). Variability in Automatic Activation as an Unobtrusive Measure of Racial Attitudes: A Bona Fide Pipeline? *Journal of Personality and Social Psychology*, 69, 1013–1027.

Grantham, T. (2004). Conceptualizing the (Dis)Unity of Science. *Philosophy of Science*, 71, 133–155.

Green, J.D., R.B. Sommerville, L.E. Nystrom, J.M. Darley, & J.D. Cohen. (2001). An fMRI Investigation of Emotional Engagement in Moral Judgment. *Science*, 293, 2105–2108.

Greenwald, A.G. & M.R. Banaji, (1995). Implicit Social Cognition: Attitudes, Self-Esteem, and Stereotypes. *Psychological Review*, 102(1), 4–27.

Harmon-Jones, E. & P. Winkielman. (2007). A Brief Overview of Social Neuroscience. In E. Harmon-Jones and P. Winkielman (eds), *Social Neuroscience: Integrating Biological and Psychological Explanations of Social Behavior*, New York: Guilford, 3–11.

Hellstrom, I.C., S.K. Dhir, J.C. Dioro, & M J. Meaney. (2012). Maternal Licking Regulates Hippocampal Glucocorticoid Receptor Transcription Through a Thyroid Hormone-Serotonin-NGFI-A Signaling Cascade. *Philosophical Transactions Royal Society of London B Biological Sciences*, 367(1601), 2495–2510.

Insel, T.R. & R.D. Fernald. (2004). How the Brain Processes Social Information: Searching for the Social Brain. *Annual Review of Neuroscience*, 27, 697–722.

Ito, T. (2010). Implicit Social Cognition: Insights from Social Neuroscience. In B. Gawronski & K. Payne (eds), *Handbook of Implicit Social Cognition* (pp. 80–92), New York: Guilford.

Ito, T. & B.D. Bartholow. (2009). The Neural Correlates of Race. *Trends in Cognitive Sciences*, 13, 524–531.

Johnson, K.J. & B.L. Fredrickson. (2005). "We All Look the Same to Me": Positive Emotions Eliminate the Own-Race Bias in Face Recognition. *Psychological Science*, 16, 875–881.

Judd, C., I.V. Blair, & K.M. Chapleau. (2004). Automatic Stereotypes vs. Automatic Prejudice: Sorting Out the Possibilities in the Payne 2001 Weapon Paradigm. *Journal of Experimental Social Psychology*, 40, 75–81.

Kubota, J.T., M.R. Banaji, & E. Phelps. (2012). The Neuroscience of Race. *Nature Neuroscience*, 15, 940–948.

LeDoux, J. (1996). *The Emotional Brain*. New York: Simon and Schuster.

Levingston, R.W. & B.B. Drwecki. (2007). Why Some Individuals Not Racially Biased? Susceptibility to Affective Conditioning Predicts Nonprejudice Toward Black. *Psychological Science*, 18(9), 816–823.

Lieberman, M.D. (2000). Intuition: A Social Cognitive Neuroscience Approach. *Psychological Bulletin*, 126(1), 109–137.

Lieberman, M.D. (2007). Social Cognitive Neuroscience. In R.F. Baumeister & K.D. Vohs (eds), *Encyclopedia of Social Psychology*. Thousand Oaks: Sage Press.

Lindquist, K. (Forthcoming). Emotions Emerge from More Basic Psychological Ingredients: A Modern Psychological Constructionist Model. *Emotion Review*.

Meyer, M.L., R.P. Spunt, E.T. Berkman, S.E. Taylor, & M.D. Lieberman. (2012). Evidence for Social Working Memory from a Parametric Functional MRI Study. *Proceedings of the National Academy of Sciences*, 109(6):1883–1888.

Mitchell, S. D. (2002). Integrative Pluralism. *Biology and Philosophy*, 17, 55–70.

Niedenthal, P.M. (2007). Embodying Emotion. *Science*, 316, 1002–1005.

Niedenthal, P., L. Barsalou, P. Winkielman, S. Krauth-Gruber, & F. Ric. (2005). Embodiment in Attitudes, Social Perception, and Emotion. *Personality and Social Psychology Review*, 9(3), 184–211.

Nosek, B.A., C.B. Hawkins & R.S. Frazier. (2011). Implicit Social Cognition: From Measures to Mechanisms. *Trends in Cognitive Sciences*, 15, 152–159.

Ochsner, K. (2007). Social Cognitive Neuroscience: Historical Development, Core Principles, and Future Promise. In A. Kruglanski & E.T. Higgins (eds), *Social Psychology: A Handbook of Basic Principles* (pp. 39–66), NY: Guilford Press.

Ochsner, K.N. & M. Lieberman (2001). The Emergence of Social Cognitive Neuroscience. *American Psychologist*, 56, 717–734.

Olsson, A., J. Ebert, M. Banaji, & E.A. Phelps. (2005). The Role of Social Groups in the Persistence of Learned Fear. *Science*, 308, 785–797.

Payne, K. & B. Gawronski. (2010). A History of Implicit Social Cognition: Where Is It Coming From? Where Is It Now? Where Is It Going? In B. Gawronski & K. Payne (eds), *Handbook of Implicit Social Cognition* (pp. 1–15), New York: Guilford.

Phelps, E. et al. (2000). Performance on Indirect Measures of Race Evaluation Predicts Amygdala Activation. *Journal of Cognitive Neuroscience*, 12(5), 729–738.

Phelps, E.A. & L.A. Thomas. (2003). Race, Behavior and the Brain: The Role of Neuroimaging in Social Behaviors. *Political Psychology*, 24(4), 747–758.

Schaller, M. & L.G. Conway. (2004). The Substance of Prejudice: Biological- and Social-Evolutionary Perspectives on Cognition, Culture and the Contents of Stereotypical Beliefs. In C.S. Crandall & M. Schaller (eds), *The Social Psychology of Prejudice: Historical and Contemporary Issues* (pp. 149–164), Lawrence: Lewinian Press.

Semin, G.R. & E.R. Smith. (2002). Interfaces of Social Psychology with Situated and Embodied Cognition. *Cognitive System Research*, 3, 385–396.

Shurick, A.A et al. (2012). Durable Effects of Cognitive Restructuring on Constructed Fear. *Emotion*, 12, 1393–1397.

Smith, E.R. & G.R. Semin, (2007). Situated Social Cognition. *Current Directions in Psychological Science*, 16, 132–135.

Squire, L.R. & J.T. Wixted. (2011). The Cognitive Neuroscience of Human Memory Since H.M. *Annual Review of Neuroscience*, 34, 259–288.

Stanley, D., E. Phelps, & M.R. Banaji. (2008). The Neural Basis of Implicit Attitudes. *Current Directions in Psychological Science*, 17(2), 164–169.

Stanley, D.A., P. Sokol-Hessner, D.S. Fareri, M.T. Delgado, M.R. Banaji & E.A. Phelps. (2012). Race and Reputation: Perceived Racial Group Trustworthiness Influences the Neural Correlates of Trust Decisions. *Philosophical Transactions of the Royal Society of London Biological Sciences*, 367, 744–753.

Wheeler, M.E. & S. Fiske. (2005). Controlling Racial Prejudice: Social-Cognitive Goals Affect Amygdala and Stereotype Activation. *Psychological Science*, 16(1), 56–63.

12
History, Causal Information, and the Neuroscience of Art: Toward a Psycho-Historical Theory

Nicolas J. Bullot

12.1 Psychological and brain theories of art appreciation

Scientists in the behavioral and brain sciences argue that experimental studies of the perceptual, hedonic, and cognitive responses to works of art are the building blocks of an emerging science of aesthetic and artistic appreciation. This science is referred to with terms such as *psychobiology of aesthetics* (Berlyne 1971), *neuroaesthetics* (Chatterjee 2011), *science of art* (Ramachandran & Hirstein 1999), or *aesthetic science* (Shimamura & Palmer 2012). Proponents of a scientific approach to art often defend a *psychological approach to art theory*. Here, I use the term 'psychological approach' broadly to denote methods that attempt to explain aesthetic and artistic phenomena by means of reference to mental and brain mechanisms. Research in the psychology of art does not essentially differ from neuroaesthetics with respect to their relations to art history and philosophical aesthetics. Both neuroscientists and psychologists tend to think that art appreciation depends on internal mechanisms that reflect the cognitive architecture of the human mind (Kreitler & Kreitler 1972; Leder et al. 2004; Zeki 1999). Like neuroscientists, psychologists present artworks as 'stimuli' in their experiments (Martindale et al. 1990; McManus et al. 1993; Locher et al. 1996). Both traditions are dominated by the psychological approach understood as an attempt to analyze the mental and neural processes involved in the appreciation of artworks. Finally, both traditions tend to identify aesthetic and artistic phenomena, and have therefore been committed to an "aesthetic-artistic confound" – see the discussion of this confound in Bullot and Reber (2013b, 164).

The study of questions in aesthetics from the standpoint of empirical psychology is often thought to begin with the work of G. Fechner (Fechner 1876) and his followers (see, e.g., Martin 1906; Pickford 1972; Berlyne 1971), who began to seek for an empirical approach to aesthetics by the end of the nineteenth century. The field of *empirical aesthetics*[1] originates from these attempts.

Numerous theories that adopt the psychological approach search for laws[2] or universals of art. Among them, D. Dutton (Dutton 2005; Dutton 2009, 51–59) and S. Pinker (Pinker 2002, 404) recently argued that there are universal signatures of art, such as virtuosity, pleasure, style, creativity, special focus, and imaginative experience. Pinker even defends the bold ahistorical claim that "regardless of what lies behind our instincts for art, those instincts bestow it with a transcendence of time, place, and culture" (Pinker 2002, 408). Zeki and Ramachandran introduce their research in neuroaesthetics as an inquiry into the ways art "obeys" the "laws of the brain" (Zeki 1999) or as a search for neurobiological laws that explain *artistic universals* (Ramachandran and Hirstein 1999).

Advocates[3] of the quest for aesthetic universals often distrust the historical methods employed in the humanities. They argue that methodologies stemming from brain and behavioral sciences should be capable of explaining art appreciation without the need to investigate the contingency and particularity (Martindale 1990, 4–7) of art-historical contexts and the appreciators'[4] *sensitivity* to such contingent art-historical contexts.

12.2   Contextualism, ahistorical theories, and the argument from the tracking of agency and functions

In contrast to the ahistorical brain theories of aesthetic universals, a number of advocates of contextualism in the humanities and social sciences of artistic behaviors employ methods that are historiographical or idiographic (Windelband 1894/1998; Lamiell 1998). I use the term *historical approach to art theory* to refer to methods based on the analysis of art-historical contexts or of the actions of intentional agents in art-historical contexts to explain artistic phenomena. In contrast to theories in psychology and neuroaesthetics that focus on the measurement of *psycho-biological* dependent variables (e.g., skin conductance, blood-oxygen-level-dependent (BOLD) contrast in fMRI), these contextualist and idiographic methods treat historical and context-specific phenomena as the most relevant variables, or *explanantia*, for explaining artistic phenomena, or *explananda*. The historical approach encompasses studies that apprehend art appreciation in the context of the history of art, styles, and particular oeuvres. It also encompasses the anthropology and sociology of art-historical contexts along with situation-specific art criticism.

According to *aesthetic contextualism* (e.g., Danto 1981; Dickie 1984/1997; Dickie 2000), theoreticians in aesthetics need to refer to historical and societal contingencies in artistic contexts (artworlds) to explain the production of art and the appreciation of particular artifacts as works of art. Contextualists often hold that a work of art is an artifact or a performance that belongs to historical categories and has properties or functions relative to particular art-historical contexts (or "artworlds"). As a historically situated artifact, the work of art is an outcome of causal interventions performed by intentional agents (e.g., an artist and a curator) embedded in a socio-cultural situation that is *historically unique* and made of unique unrepeatable events and irreplaceable objects. Contextualists investigate the consequences of this historical embedding for the theory of identity, appreciation, understanding, and evaluation of works of art. The premises of contextualism can lead historicists to assume the truth of relativism about artistic values, a position which can conflict with the belief in the existence of psychological universals defended by the brain theories of art appreciation. In contrast to the aesthetics of *disinterestedness*, which contends that appreciating an art object implies a "psychic distance" from functions of the art object, contextualism suggests that scholars need to develop a cognitively rich account of aesthetic appreciation of artistic functions, which takes the richness of our contextual knowledge about artworks as an essential condition of their appreciation.

Proponents of contextualism stress the importance of historical factors and context-specificity in art and its appreciation. Consequently, their view predicts that scholars who seek to explain art appreciation need to investigate art appreciators' *sensitivity to* art-historical contexts. This prediction seems to conflict with the principles of neuroaesthetics and the psychological approach. Because numerous defenders of the psychological approach have often investigated art appreciation without analyzing the appreciator's sensitivity to art-historical contexts, a number of scholars who adopt contextualism express doubts that psychological and neuroscientific theories can succeed in explaining art appreciation.[5]

Contextualists have discussed a variety of objections against a purely psychological approach to human behavior, including methodological objections derived from research on the cross-cultural variability of human behavior (Henrich, Heine & Norenzayan, 2010). In the context of the science of art, Bullot and Reber (2013a, 125) propose a basic objection aimed at psychological theories of artistic behavior, which can be formulated in terms of sensitivity to context-specific functions. I will call this objection the *argument from the tracking of artistic agency and functions*; it may be expressed as follows:

First, the appreciator's competence in artistic appreciation of a work of art is a form of *sensitivity to* – or, an ability *to track* – characteristics of the art-historical context of this work such as the context-specific functions of the work and the agency of its maker. Second, most psychological and neuroscientific theories do not explain the appreciator's ability to track characteristics of the art-historical context of the work. Therefore, most psychological and neuroscientific theories do not explain the appreciator's artistic appreciation.

This objection seems sound when directed at ahistorical theories in the behavioral and brain sciences that investigate the psychological or neural responses to art without a theory of the neural basis of the appreciator's sensitivity to art-historical contexts. Studies of that sort include works in empirical aesthetics[6] that use reproductions of artworks (prototypically, paintings) as stimuli without manipulating the participants' knowledge of the art-historical contexts relevant to appreciation of the works presented as stimuli. While an increasing number of such psychological studies reveal interesting phenomena, they tend to use a way of displaying the artwork in which the historical context of the work, its functions, and the agency of its creator are not relevant, or cannot be assessed by the participant in the experimental study.

12.3   A psycho-historical theory of art appreciation

As an alternative to radical forms of psychologism or historicism, I have proposed a theory of art appreciation that combines historical contextualism and psychology: the *psycho-historical theory of art appreciation* ('psycho-historical theory' henceforth) – see Bullot (2009), Bullot and Reber (2013a, 2013b). In Sections 12.4 and 12.5, I outline the principles and hypotheses of this theory. These elements aim to complement the ahistorical methods that tend to dominate the theories of art production and appreciation in the brain in behavioral sciences. In contrast to acontextual theories, the psycho-historical account incorporates contributions and insights from both contextualism and the historical approach, and aims at providing contextualist principles that can serve the development of experimental research within the psychological approach.

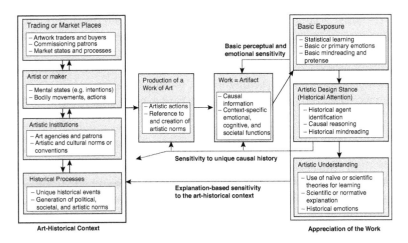

Figure 12.1   The psycho-historical theory of art appreciation

Solid arrows indicate relations of causal and historical generation. Broken arrows indicate information-processing and representational states of the appreciators' mind that refer back to earlier historical stages in the production and transmission of a work.

Source: Adapted from Bullot and Reber 2013a.

Figure 12.1 depicts the historical concepts of the theory and their relations (Section 12.4) and the appreciation of the work through three modes of information processing (Section 12.5).

12.4   Historical principles

12.4.1   Principle of the art-historical context

An art-historical context encompasses agents and processes that causally intervene in the making and transmission of art. Here, I use the term *art-historical context* to refer to the complex of human agents, artistic and trading institutions, and historical and political processes that govern the production of artistic artifacts and behaviors along with their subsequent evaluation, trade, reproduction, and conservation. Artists, patrons, sellers, politicians, and audiences are intentional agents who build this context through deployment of their individual and collective agency. Scholars have studied art-historical contexts from a variety of perspectives. Contextualist philosophers like Danto and Dickie, who use the concept of *artworld* to refer to art-historical contexts, have persuasively argued that appreciators cannot understand the artistic identity and values of artworks without being sensitive to their art-historical contexts (artworlds) (Danto 1964, 1981; Dickie 2000, 1984/1997). Art historians also investigate art-historical contexts in order to understand the oeuvres of particular artists or the style of particular schools. Similarly, anthropologists or sociologists (e.g., Bourdieu 1979/1987; Tanner, ed. 2003; Crane 1989) use the methods of social sciences to examine trends or mechanisms in art-historical contexts to explain artistic *explananda*.

12.4.2 Principle of artifact functions

A work of art is an artifact that has context-specific functions. When asserting that the work of art is an artifact (see Figure 12.1), I use the term *artifact* in a broad sense that refers to an object or a performance intentionally brought into existence through causal intervention of human action and intentionality (Margolis and Laurence 2007; Hilpinen 2011). This usage entails that all artistic performances are artifacts in the minimal sense of being products of human actions. Artifacts in this sense have functions (Millikan 1984; Dennett 1990; Dennett 1987). Reference to an original context and function is not always sufficient to explain the variety of functions that artifacts have because many artifacts acquire novel functions or have prominent functions abandoned over time. Though several alternative accounts of artifact functions have been proposed, it is significant that all the accounts I studied refer to the *historical context* of artifacts as a means to explain how they acquire proper or accidental *functions*. Referring to particular historical contexts is indispensable in elucidating the functions of artifacts because many of such functions are context-specific, and knowing the historical context of an artifact can help investigators elucidate its functions.

An illustration of the historicality of artifact functions is provided by Danto, who has argued that one of the functions or roles of Duchamp's readymades and Warhol's Brillo Boxes is to bring to appreciators' reflexive awareness an understanding of what art is, and that such function is a product of the history of Modernism (Danto 1998; Danto 2009). If Danto's interpretation is correct, this function is context-specific because it is a reproduced psychological effect (the conscious awareness of a question) that cannot take place in absence of an adequate art-historical context and understanding of the prevalent norms in such context. This point most clearly holds with respect to readymades, but it could be transposed to any artistic functions. If artistic functions are context-specific, as predicted by the contextualist approach, this point would apply to all functions that works of art gain or lose over historical time. In agreement with Parsons and Carlson (2008) and other research on artifact,[7] it seems correct to conclude that artifact and artwork functions need to be explained by historical accounts that take into consideration context-specific situations and norms.

12.4.3 Principle of causal-historical information

The work of art is a carrier of causal and historical information about agents, agency, and other elements in art-historical contexts. As stressed by Berlyne and other psychologists, it is useful to adopt an information-theoretic conception of the work of art to account for the different sources of knowledge that its appreciation can provide (Berlyne, ed. 1974, 6–8). Berlyne noted that the characteristics or features of an artwork can be sources of syntactic, cultural, expressive, and semantic information. However, Berlyne's conception is missing an important aspect because it is *ahistorical*. Specifically, it does not make clear that the information carried by a work is the end product of a causal history and the way appreciators extract information about the work is driven by inquiries about the past. Art appreciation is driven by the appreciator's sensitivity to the present states and the causal history of observable properties of the work. This causal history is schematized in Figure 12.1. In the psycho-historical account, this causal-historical aspect can be captured by conceiving of each of the intrinsic features of a work as a *trace*[8] or *carrier of causal information* that can be processed to acquire knowledge about relevant art-historical contexts. In what follows, I will use the term *causal information* to refer to objective and observer-independent causal relations that affect the mental processing of the artwork and can be retrieved by means of causal reasoning.[9]

As depicted in Figure 12.1, any artifact or work of art carries causal information, or causal data. For instance, each hand stencil or imprint in the Gallery of Hands at the Chauvet cave (Chauvet-Pont-d'Arc, France; see Clottes 2005) is a trace that carries causal information because it is a set of facts that are reliable pieces of evidence of antecedent events (dating from the Paleolithic era), such as a person blowing pigments on a hand placed against the cave wall. In virtue of the nature of traces and causal information, certain properties such as the location and chemical composition of paint pigments connect the present facts (e.g., this *stencil's outlining a hand*) to antecedent events (e.g., *somebody's blowing these paint pigments onto a hand*) because they are connected through a causal chain. Given the ubiquitous nature of causal information, this point can be generalized to any artifacts, artworks, or art forms. For instance, the celluloid films used to project *Battleship Potemkin* (1925) by Sergei Eisenstein on cinema screens are carriers of traces that can inform about the scenes or actors filmed or about the decisions made by the director.

Written works like poems and novels, and sound recordings, are carriers of information about past art-historical contexts too, and so forth for any artistic media.

12.5   Psychological hypotheses

When appreciators perceive a work or some of its representations, they are exposed to the causal-historical information it carries. The psycho-historical theory suggests three basic hypotheses about the processing of this causal-historical information.

12.5.1   Basic exposure

An elementary mode of appreciation is *basic exposure* to the work or one of its reproductions (see right side of Figure 12.1). Basic exposure is the variety of processes that spontaneously occur when appreciators perceive an artwork without having knowledge about its causal history and art-historical context. The direct experience of a work of art depends on a variety of processes and biases of perception, focused awareness and attention, illusion, memory systems and recollection, imagination, symbolic interpretation, and exploratory action. Many of such processes are necessary conditions of the appreciators' exposure to and immediate preferences for particular works of art (and such preferences that might not derive from inferences: see, e.g., Zajonc 1980). As suggested by evolu-tionary accounts, appreciators' immediate preferences might exhibit universal aesthetic biases, such as symmetry and sexual dimorphism in faces (Rhodes 2006). If these accounts are correct, such universal biases should be manifested in basic exposure.

Instead of attempting a comprehensive exploration of these numerous basic processes and biases, I will focus on three examples: (1) the implicit learning of regularities and style; (2) the elicitation of emotions; and (3) pretense. Although such processes may involve aesthetic pleasures, they are historically shallow because they do not provide the appreciator with explicit knowledge of the connections between the work and its original art-historical context.

(1) Since artworks carry causal-historical information, repeated expo-sure to a work may nonetheless allow its appreciators to *implicitly* develop their sensitivity to historical facts or rules, even if such appreciators are deprived of knowledge about the original artistic context. Perceptual exposure to a work of art leads to types of learning that may occur even if the learner does not possess any

explicit knowledge about the history of the work. They may lead to implicit forms of causal-historical learning of style. However, there is reason to doubt that artistic understanding is a prerequisite of basic stylistic classifications. For instance, pigeons can learn to classify artworks according to stylistic features, and it seems unlikely that such learning demonstrates an understanding of art (Watanabe et al. 1995). This rather suggests that discrimination of basic styles stems from probabilistic learning through basic exposure that does not require an understanding of the causal processes underlying styles of individual artists (Goodman 1978) or of historical periods (Arnheim 1981; Wölfflin 1920/1950; Panofsky 1995). Such understanding is more likely to derive from inferences based on historical theories rather than similarity.[10]

(2) Basic exposure can also be specified with respect to *emotions*. The sensory exposure to form and content of a work of art can elicit a variety of non-historical emotional responses.[11] These may include the emotions (Robinson 2005, see sections 5 and 7) that are sometimes described as *basic* (Ekman 1992) or *primary* (Damasio 1994) – such as anger, fear (Robinson 1995; Walton 1978), disgust, and sadness – and other basic responses such as 'startle', erotic desire (Freedberg 1989), enjoyment, or feeling of empathetic engagement (Freedberg and Gallese 2007). The historical knowledge that appreciators gain from the elicitation of these basic emotions through basic exposure to a work is shallow at best.

(3) The appreciator's basic perceptual exposure to the work can prompt processes aimed at representing mental states, so-called *mindreading* (Carruthers 2009; Nichols and Stich 2003). Philosophical arguments presented by Walton (1990, 1973), Currie (1995), and Nichols (2006) provide reason to think that mindreading is essential to art appreciation. This is because, in responding to a work of art in the mode of basic exposure, appreciators often elaborate on free imaginative games, and on pretense involving the attribution of fictional beliefs or desires to characters. Though these games may be stunning constructions of the appreciators' imagination (Nichols and Stich 2003; Harris 2000), they do not entail sensitivity to the causal history of artworks.

Basic perceptual exposure to artworks is the mode of art appreciation most frequently investigated by advocates of the psychological approach. The psycho-historical theory suggests that this approach is too narrow to provide a comprehensive account of art appreciation. As demonstrated by the argument from the tracking of agency and functions in art-historical contexts (Section 12.2), research restricted to basic exposure cannot account for the artistic understanding that stems from the appreciators' sensitivity to art-historical contexts and to art functions in such contexts. A distinction between processing of artworks at the level of basic exposure and at the level of causal-historical understanding is necessary to develop a comprehensive theory. This kind of understanding requires the adoption of an 'artistic design stance.'

12.5.2   The artistic design stance

Once appreciators are exposed to an artwork, they may engage in an investigation of the causal history that has led to the production and transmission of the work understood as an exemplar that can be traced over time. This mode of appreciation may be termed the *artistic design stance* (see Figure 12.1). The concept of 'design stance' has been introduced by Dennett, who used it to denote a strategy aimed at explaining and predicting the behavior of an artifact by means of reference to its mechanisms and functions (Dennett 1978; Dennett 1971, 87–88). In contrast to basic exposure, this mode of appreciation is no longer historically shallow because it relies on the processing of information as related to past causal facts. It is a process by means of which appreciators become sensitive to, and track over time, artistic agency and functions in the art-historical context of the work.

Empirical research conducted and reviewed by Kelemen and Carey (2007) supports the hypothesis that our understanding of artifact concepts relies on the adoption of a *design stance*. Strategies like the design stance must be adopted when appreciators reason about artifacts and attempt to understand their functions. Given the fact that works of art are artifacts and products of human actions, appreciators of art should also adopt a strategy like the design stance.

What are the characteristics of an *artistic* design stance? Dennett does not address this question. If the design stance is a means whereby appreciators develop artistic understanding, we can conjecture that the artistic design stance should involve at least three kinds of processes (Figure 12.1); only the third one would be exclusive to art appreciation. (1) In adopting a *causal stance*, art appreciators would inquire into causal information to support the premises of their reasoning about the causal origins of the observable properties of the work. (2) In adopting an *individuation stance*, appreciators would expand the causal stance to refine hypotheses about the unique causal history of the work understood as unique exemplar, the genealogy of its functions, and the particular agents who produced it. (3) In adopting an *intentional stance*, apprecia-tors would use their mindreading abilities to establish that the artwork under consideration was designed to meet context-specific *artistic* inten-tions and artistic aims.

(1) Works of art carry diverse sorts of information. When an appreciator begins to infer from observable features of the work the causal history of unobserved actions that have led to these observable features, this appreciator begins to engage in the *causal stance* necessary to the design stance. This claim is supported by evidence that humans spontaneously try to track down the cause of an event especially if the event is surprising or salient – a search for causes that triggers causal reasoning (Heider 1958; Gelman 2003, chapter 5). Because the exhibit of artworks often confers on them the status of salient beautiful, majestic, surprising, intriguing, thought-provoking, or even offending (Julius 2002) objects, appreciators who direct their attention to them will be led to a search for causal and historical information.

(2) Once appreciators have begun to adopt the causal stance, they are in a position to process information as causal information and adopt an *individuation stance* in an attempt to answer the questions that need to be addressed in order to determine authorship of the work and causal processes that underlie its nature and identity. For instance, who were the persons involved in the making of the work? When and where was this work made and transmitted? What are its boundaries in space and time? The ability to *reliably* answer these ontological questions has its roots in an ability to track the history and distinctness of the artwork over repeated exposure by means of inferences and theory-based reasoning. Identification and locali-zation of agents and events important to the history of the work can be refined through the careful study of the work, with the help of different methods or by means of identifying narratives[12] with differences of causal and explanatory depth. For instance, one may identify and track an artwork by means of perceptual recognition, analysis of its chemical constituents (Coremans 1949), scrutiny of its stylistic features, and, more generally, abductive inferences about its conditions of production. These are epistemic processes that inves-tigate clues to the identity of a work. Appreciators need to enact such processes to generate veridical narratives about a work, such as narratives that authenticate genuine works and distinguish them from forgeries and lookalikes.

(3) In addition to triggering causal attribution, the adoption of the causal and individuation stances may also prompt mindreading, and therefore lead to an artistic *intentional stance* (Dennett 1987). In basic exposure, appreciators use mindreading to represent the mindreading-related content of a work without investigating its art-historical context. In contrast, the artistic intentional stance leads appreciators to inquire into the mental states of important agents in the original art-historical context of the work (e.g., intentions of the artist or patron). Appreciators may use a wide variety of means to engage in the artistic intentional stance. They might use simulation (Goldman 2006) or reasoning based on relevance and optimality (Sperber and Wilson 2002; Dennett 1990) to interpret the intentions of agents in bygone art-historical contexts. For example, appreciators may interpret an artist's intention as aimed at producing a work the function of which is to elicit a specific kind of emotion or cognitive process in appreciators' minds. In this way, mindreading driven by the intentional stance can enable audiences to develop a design stance aimed at apprehending an artwork from the perspective of the artist.

In contrast to basic exposure, where the audience focuses on observable aspects of the finished product, an appreciator who takes the design stance can also imagine alternative solutions to the artistic problem and hence use imagination, counterfactual reasoning (Nichols and Stich 2003), or thought experiments to infer the way the artist might have solved the artistic problem addressed by the work. This kind of mindreading would refine an appreciator's sensitivity to the causal history of the work and might therefore enable artistic understanding (see Figure 12.1).

12.5.3   Artistic understanding

If appreciators of a work take the design stance for guiding their interpretation of the work, they will increase their *sensitivity to* and *proficiency with* the art-historical context and contents of this work. This increase in proficiency enables a third mode of art appreciation, *artistic understanding*. Appreciators have *artistic understanding* of a work if art-historical knowledge, acquired as an outcome of the design stance, provides them with an ability to explain the artistic status and significance of the work.

Because there are multiple forms of understanding and explanation (Keil and Wilson 2000; Keil 2006; Ruben 1990), we need a pluralistic approach to artistic understanding. At a minimum, there is a need to distinguish two modes of artistic understanding. The *normative mode* of artistic understanding aims to identify and evaluate the artistic merits of a work, and more generally its value. It is often based on contrastive explanations that compare the respective art-historical values of sets of artifacts. These evaluations are often viewed as essential to the practice of art critics and art historians. The *scientific mode* of artistic understanding does not aim to provide normative assessments, but to explain artistic *explananda* with the methods and approaches of specific scientific traditions. In a way that parallels the combination of normative and scientific aspects in folk-psychology, the normative and scientific modes of understanding are often intermingled in common sense thinking about art and scholarly writings about art (Berlyne 1971, 21–23; Roskill 1976/1989).

12.6 The advantages of adopting a psycho-historical framework

Thus far, I have outlined basic components of a psycho-historical theory of art appreciation, which encompasses both historical principles (Section 12.4) and psychological hypotheses (Section 12.5). In this Section, I will consider the psycho-historical theory as a whole and discuss its main advantages over its psychological and historical competitors.

12.6.1 Unification

The primary advantage is *unification*. The psycho-historical theory is novel because it is the first proposal that apprehends the empirical psychology of artwork processing and the historical research on art and its appreciation as two inseparable components of the same theoretic and scientific endeavor. The primary advantage of the theory is thus to integrate the two main streams of research on art and its appreciation.

In accord with the historical approach, the psycho-historical theory adopts the contextualist insight that art appreciation is deployed by the appreciator's sensitivity to causal and historical properties of art-historical contexts. However, the psycho-historical theory also adopts the psychological and neuroscientific view that appreciation can be explained by means of an information-processing theory based on the empirical investigation of emotions and cognitive mechanisms of the brain. In contrast to many historical works that just ignore psychological processing or use the analytical resources of common-sense psychology, the psycho-historical theory is naturalized in the sense that it is constrained and informed by empirical research in psychology, neuroscience, and cognitive science. For instance, the reference to the design stance and its components is compatible with empirical research on psychological essentialism. Furthermore, in contrast to psychological studies that investigate universal and innate aesthetic responses while ignoring the role of historical processes (e.g., cultural norms) and historical knowledge in art appreciation, the psycho-historical approach takes into account the fundamental historicity of art and its appreciation.

Another related advantage of the psycho-historical theory is that it does not apprehend artifact and art appreciation from the standpoint of a single brain mechanism or module in isolation from other cognitive systems. Numerous theories in the psychological approach are partitioned on the basis of links between components or modules of human cognitive architecture and certain artistic genres. For instance, some theories focus on the paired study of vision and the visual arts (Pickford 1972; Arnheim 1956/1974; Zeki 1999), audition and music processing (Peretz 2006), expressive artworks and emotional responses, or imagination and pretense. This way of using the architecture of the mind to partition art appreciation leads to fragmentation of the research in aesthetics and cannot provide a comprehensive theory of artistic behavior. In contrast to this tradition, the psycho-historical theory defends a cognitively rich account of artistic appreciation based on the view that art appreciation involves a variegated set of interacting psychological processes.

12.6.2   Accounting for the narrative and causal brain processes

The theory presents more specific advantages. By making processes of historical inquiry an integral part of the processes performed in response to the appreciators' exposure to art, the psycho-historical theory accounts for the role of narrative or historical explanations and theories in art appreciation. The latter has been neglected in aesthetics. According to the psycho-historical theory, the appreciator's understanding of a work has to rely on naïve or scientific theories (Gopnik and Meltzoff 1997; Kelemen and Carey 2007; Murphy and Medin 1985) and causal reasoning. As noted by Gopnik and Meltzoff (1997, 34–41), theories have structural characteristics such as conceptual coherence, power of generalization, and representations of causal structures. In the context of art appreciation, these characteristics allow inquirers who use theories to make predictions, produce cognitively rich interpretations of an artwork, and generate abductive inferences about art-historical contexts (on abductive inference (inference to the best explanation), see, e.g., Carruthers 2006; Lipton 1991/2004). Theories of the art-historical context appear necessary to the appreciator's competence in reliably identifying and explaining aesthetic properties such as style, genre, and context-dependent meaning or function.

Consider again style. I noted that basic exposure might lead appreciators to recognize artistic styles through probabilistic learning and similarity-based classifications. Since such processing remains shallow with regard to historical processing, basic exposure can hardly provide appreciators with explanations and justifications for the identification of styles and their similarities. In contrast, appreciators who develop a robust form of artistic understanding can use their historical knowledge to guide their inquiries into relevant aspects of art-historical contexts and identify stylistic properties more reliably. Naïve or scientific theories are helpful, if not indispensable in this case, because stylistic properties of individuals or schools are notoriously difficult to identify. When reference to style is made to classify art schools and identify periods in art history,[13] the justification of such periodization must appeal to theories of art-historical contexts because only historical knowledge can provide explanations for such classifications.

Theories of aspects of an art-historical context can also inform mind-reading of important artistic intentional agents in the given context. This can be illustrated by the role of theories to inform perspective-taking or simulations aimed at understanding the decisions made by an artist or attempting to re-enact the artist's decision or experience (Croce 1902/1909; Croce 1921).

If the psycho-historical account is correct in predicting the centrality of the design stance, this prediction opens up the possibility of errors and misunderstanding in art interpretation, which may depend on fallacies or incorrect explanation of the relationships between the work and its art-historical context, and not just on mistakes in the processing of observable features of the artwork (in basic exposure). To my knowledge, the epistemology of such errors has never been investigated.

12.6.3   Accounting for epistemic and historical emotions in artistic understanding

Another advantage of the psycho-historical theory is that it provides a novel framework to study the consequence of the design stance on emotions related to art appreciation. Inferences about the causes of an artwork due to the design stance may trigger emotions like any other epistemic evaluations. Though emotions are often automatically elicited through basic exposure, the design stance and artistic understanding can elicit different kinds of emotions.

Several theorists distinguish between a primary and a secondary appraisal of the stimulus that has elicited the emotion experience (Lambie and Marcel 2002; Frijda 1986, 186–192). Arguably, many of the emotions that appreciators experience after adopting the design stance are secondary emotions or feelings. On the basis of such secondary emotions, appreciators could achieve a better 'emotional understanding' of the content of a work and improve their knowledge of the historical context and the artist's intention. The quality of the emotion elicited by an artwork will depend on the kind of causal attribution or appraisal made to agents in relation to the art-historical context. In other terms, the same artwork may elicit different emotions, depending on the causal attributions the audience makes. Consider again the argument from the tracking of function and Warhol's Brillo Boxes. The Brillo Boxes are appreciated from a different perspective by someone who ignores the artworld of modernism (Danto 1998, Clark 2001; Greenberg 1961; Fried 1998) in Western art and someone who has historical proficiency with Warhol's art-historical context (e.g., Warhol's relationship to Marcel Duchamp and the artists who founded Pop Art). Arguably, in comparison to basic exposure, taking the design stance should result in improved artistic understanding and experience because the design stance enables appreciators to draw causal inferences resulting in the experience of a wider range of emotions associated with historical understanding.

12.6.4   Accounting for the tracking of uniqueness in art appreciation

As a theory that aims to explain how appreciators track aspects of art-historical contexts over time, the psycho-historical theory has also the advantage that it provides a novel way to address the classic conundrum of the artistic appreciation of look-alikes (Danto 1981) and forgeries (Dutton 1979). If art were appreciated only at the level of basic exposure, and thus without causal understanding, two artworks that look exactly alike should elicit equivalent responses in appreciation. The *Brillo Boxes* made by Warhol (Danto 2009) would elicit equivalent appraisal as the stacks of Brillo boxes in supermarkets. However, analysis of the artistic appreciation of look-alikes and historical records of responses to the discovery of forgeries contradict the prediction of an equivalent appraisal of look-alikes. Appreciators value look-alikes differently once they understand that they have a different causal history. Consider the classic example of van Meegeren's forgeries. The discovery that works allegedly painted by Vermeer were in fact fabricated by van Meegeren has led their audience to revise their artistic value because the causal history of the works, and their relations to their maker and art-historical context, matter to their artistic value. Van Meegeren's forgeries are misleading when they are taken to be material evidence of Vermeer's past action and artistry. The common preference for originals compared to indiscernible replicas or forgeries is inexplicable from a psychological approach that considers only basic exposure. This reasoning is another line that supports the psycho-historical account.

12.7 Recapitulation

The combination of my analysis of the art-historical context and the production and transmission of the artwork (Section 12.4) and of the three modes of art appreciation (Sections 12.5) composes the psycho-historical theory of art appreciation. The core hypothesis of the theory is that appreciators' response to artworks varies as a function of their sensitivity to relevant art-historical contexts. The more appreciators are proficient with the relevant art-historical context of a work as a result of adopting the design stance, the greater the chance that they develop an adequate artistic understanding of this work. According to the theory, the adoption of the design stance is the core process that underlies the development of artistic learning and of both normative and scientific modes of artistic understanding. The theory predicts that if appreciators of a work adopt the design stance and subsequently learn about relevant art-historical contexts, then they will increase their ability to interpreting and understanding the work. This prediction contradicts the claim made by universalists (Fodor 1993; Pinker 2002; Ramachandran 2001) adopting the psychological approach that the appreciation of a work does not require knowledge of the art-historical context.[14]

Notes

1. See Berlyne 1971, 1974; Martindale et al. 1990; Martindale 1984, 1990; Locher 2003; Pickford 1972; Kreitler and Kreitler 1972.
2. For a defense of the search for laws in empirical aesthetics, see Martindale 1990, 3–13.
3. See, among others, Martindale 1990; Ramachandran 2001.
4. The term *appreciator* refers to the person who is making the appreciation, regardless of whether this person is the artist or the member of an audience.
5. Doubts are expressed in these works: Currie 2004; Gombrich 2000; Dickie 2000; Parsons 1987; Margolis 1980; Dickie 1962.
6. See, e.g., Martindale et al. 1990; Locher et al. 1996; Kawabata and Zeki 2004; Vartanian and Goel 2004.
7. Kelemen and Carey 2007; Kelemen 1999; German and Johnson 2002; Matan and Carey 2001.

8. Here, the term *trace* is used to denote causal information about the past. It is important to distinguish the information-theoretic concepts of traces and causal information from the common-sense concept of trace that refers to visible marks or stains on a surface. The extension of the former is much broader than that of the latter. Consequently, the information-theoretic concepts apply to all art forms. Any particular property or fact that conveys causal information about the past properties or facts is a trace. For any given work of art of any genre, the existence of this work depends on the existence of characteristics that inform about the work because they are traces that convey information about the work.

9. The term *information* combined with a qualifier such as *causal* or *natural* denotes observer-independent regularities or laws between physical phenomena or facts. See, e.g., Dretske 1988; Millikan 2000; Millikan 2004, p. 33.

10. Murphy and Medin 1985; Arnheim 1981; Wölfflin 1920/1950; Panofsky 1995.

11. Ducasse 1964; Peretz 2006, 22–23; Robinson 2005; Robinson 1995.

12. On the role of identifying narratives in conceptual art, see Davies 2007.

13. Goodman 1978; Walton 1987. Wölfflin 1920/1950; Riegl 1890/1992; Gombrich 1979, chapter 8.

14. See Fodor 1993, 51–53. Fodor argues that art appreciation is not necessarily based on the understanding of historical (real) intentions because appreciators' interpretation of a work could derive from the imaginings of virtual artistic intention. Although imagining virtual intentions may facilitate an audience's understanding of a work, the psycho-historical framework predicts that adopting the design stance for the acquisition of knowledge about historical intentions facilitates understanding and enhances appreciation more than virtual intentions.

References

Arnheim, R. (1956/1974). *Art and Visual Perception. A Psychology of the Creative Eye.* Berkeley, CA: University of California Press.

Benjamin, W. (1936/2008). The Work of Art in the Age of Its Technological Reproducibility, Second Version, trans. E. Jephcott, R. Livingstone & H. Eiland. In *The Work of Art in the Age of Its Technological Reproducibility, and Other Writings on Media,* ed. M. W. Jennings, B. Doherty & T. Y. Levin (pp. 19–55). Cambridge, Mass.: Harvard University Press.

Berlyne, D. E. (1971). *Aesthetics and Psychobiology.* New York: Meredith Corporation.

Berlyne, D. E. (Ed.). (1974). *Studies in the New Experimental Aesthetics: Steps toward an Objective Psychology of Aesthetic Appreciation.* Washington, DC: Hemisphere Publishing Corporation.

Bourdieu, P. (1979/1987). *Distinction: A Social Critique of the Judgment of Taste,* trans. R. Nice. Cambridge, Mass.: Harvard University Press.

Bullot, N. J. (2009). Material Anamnesis and the Prompting of Aesthetic Worlds: The Psycho-Historical Theory of Artworks. *Journal of Consciousness Studies,* 16(1), 85–109.

Bullot, N. J. and R. Reber (2013a). The Artful Mind Meets Art History: Toward a Psycho-historical Framework for the Science of Art Appreciation. *Behavioral and Brain Sciences*, 36(02), 123–137.

Bullot, N. J. and R. Reber (2013b). A Psycho-historical Research Program for the Integrative Science of Art. *Behavioral and Brain Sciences*, 36(2), 163–180.

Carruthers, P. (2006). *The Architecture of the Mind: Massive Modularity and the Flexibility of Thought*. Oxford: Oxford University Press.

Carruthers, P. (2009). Mindreading Underlies Metacognition. *Behavioral and Brain Sciences*, 32(2), 164–182.

Chatterjee, A. (2011). Neuroaesthetics: A Coming of Age Story. *Journal of Cognitive Neuroscience*, 23(1), 53–62.

Clark, T. J. (2001). *Farewell to an Idea: Episodes from a History of Modernism*. New Haven: Yale University Press.

Clottes, J. (2005). *Return to Chauvet Cave: Excavating the Birthplace of Art – The First Full Report*. London: Thames & Hudson.

Coremans, P. B. (1949). *Van Meegeren's Faked Vermeers and De Hooghs: A Scientific Examination*. Amsterdam: J. M. Meulenhoff.

Crane, D. (1989). *The Transformation of the Avant-Garde: The New York Art World, 1940–1985*. Chicago: University of Chicago Press.

Croce, B. (1902/1909). *Aesthetic as Science of Expression and General Linguistic* [1902], trans. D. Ainslie. London: Macmillan and Co.

Croce, B. (1921). *The Essence of Aesthetic*, trans. D. Ainslie. London: William Heinemann.

Currie, G. (1995). Imagination and Simulation: Aesthetics Meets Cognitive Science. In *Mental Simulation: Evaluations and Applications*, eds. M. Davies and T. Stone (pp. 151–169). Oxford: Blackwell.

Currie, G. (2004). *Arts and Minds*. Oxford: Oxford University Press.

Damasio, A. (1994). *Descartes' Error: Emotion, Reason, and the Human Brain*. New York: G. P. Putnam.

Danto, A. C. (1964). The Artworld. *Journal of Philosophy*, 61, 571–584.

Danto, A. C. (1981). *The Transfiguration of the Commonplace: A Philosophy of Art*. Cambridge, Mass.: Harvard University Press.

Danto, A. C. (1998). *After the End of Art: Contemporary Art and the Pale of History*. Princeton: Princeton University Press.

Danto, A. C. (2009). *Andy Warhol*. New Haven: Yale University Press.

Davies, D. (2007). Telling Pictures: the Place of Narrative in Late Modern "Visual Art." In *Philosophy and Conceptual Art*, eds. P. Goldie and E. Schellekens (pp. 138–156). Oxford: Oxford University Press.

Dennett, D. C. (1971). Intentional Systems. *The Journal of Philosophy,* 68(4), 87–106.

Dennett, D. C. (1978). *Brainstorms*. Montgomery: Bradford Books.

Dennett, D. C. (1987). *The Intentional Stance*. Cambridge, Mass.: MIT Press.

Dennett, D. C. (1990). The Interpretation of Texts, People and Other Artifacts. *Philosophy and Phenomenological Research,* 50 (Issue Supplement), 177–194.

Dickie, G. (1962). Is Psychology Relevant to Aesthetics? *The Philosophical Review*, 71(3), 285–302.

Dickie, G. (1984/1997). *The Art Circle: A Theory of Art* [1984]. Evanston, IL: Chicago Spectrum Press.

Dickie, G. (2000). The Institutional Theory of Art. In N. Carroll (Ed.), *Theories of Art Today* (pp. 93–108). Madison, Wisconsin: The University of Wisconsin Press.

Dretske, F. (1988). *Explaining Behavior: Reasons in a World of Causes.* Cambridge, Mass.: MIT Press.

Ducasse, C. J. (1964). Art and the Language of the Emotions. *The Journal of Aesthetics and Art Criticism*, 23(1), 109–112.

Dutton, D. (1979). Artistic Crimes: The Concept of Forgery in the Arts. *British Journal of Aesthetics*, 19(4), 302–341.

Dutton, D. (2005). Aesthetic Universals. In *The Routledge Companion to Aesthetics*, 2nd edn, ed. B. N. Gaut & D. M. Lopes (pp. 279–292). London: Routledge.

Dutton, D. (2009). *The Art Instinct: Beauty, Pleasure & Human Evolution.* Oxford: Oxford University Press.

Ekman, P. (1992). An Argument for Basic Emotions. *Cognition & Emotion*, 6(3–4), 169–200.

Evans, G. (1982). *The Varieties of Reference.* Oxford: Oxford University Press.

Fechner, G.T. (1876). *Vorschule der Aesthetik [Elements of Aesthetics].* Leipzig: Druck und Verlag von Breitkopf & Härtel.

Fodor, J. A. (1993). Déjà Vu All Over Again: How Danto's Aesthetics Recapitulates the Philosophy of Mind. In *Danto and His Critics*, ed. M. Rollins (pp. 41–54). London: Blackwell.

Freedberg, D. (1989). *The Power of Images: Studies in the History and Theory of Response.* Chicago: University of Chicago Press.

Freedberg, D. & Gallese, V. (2007). Motion, Emotion and Empathy in Esthetic Experience. *Trends in Cognitive Sciences*, 11(5), 197–203.

Fried, M. (1998). *Art and Objecthood: Essays and Reviews.* Chicago: University of Chicago Press.

Frijda, N. H. (1986). *The Emotions.* Cambridge/ Paris: Cambridge University Press/ Editions de la Maisons des Sciences de l'Homme.

Gelman, S. A. (2003). *The Essential Child: Origins of Essentialism in Everyday Thought.* Oxford: Oxford University Press.

German, T. P. & Johnson, S. C. (2002). Function and the Origins of the Design Stance. *Journal of Cognition and Development*, 3(3), 279–300.

Goldman, A. I. (2006). *Simulating Minds: The Philosophy, Psychology, and Neuroscience of Mindreading.* Oxford: Oxford University Press.

Gombrich, E. H. (1979). *The Sense of Order: A Study in the Psychology of Decorative Art.* London: Phaidon.

Gombrich, E. H. (2000). Concerning "The science of art": Commentary on Ramachandran and Hirstein. *Journal of Consciousness Studies*, 7(8–9), 17.

Goodman, N. (1978). *Ways of Worldmaking.* Indianapolis: Hackett.

Gopnik, A. and Meltzoff, A. N. (1997). *Words, Thoughts, and Theories.* Cambridge, Mass.: MIT Press.

Greenberg, C. (1961). *Art and Culture.* Boston: Beacon Press.

Harris, P. L. (2000). *The Work of the Imagination.* Malden, Mass.: Blackwell.

Heider, F. (1958). *The Psychology of Interpersonal Relations.* New York: Wiley.

Henrich, J., Heine, S. J. & Norenzayan, A. (2010). The Weirdest People in the World? *Behavioral and Brain Sciences* 33(2–3): 61–83.

Hilpinen, R. (2011). Artifact (revised version). *Stanford Encyclopedia of Philosophy*, ed. Edward Zalta. http://plato.stanford.edu/entries/artifact/

Julius, A. (2002). *Transgressions: The Offences of Art*. London: Thames & Hudson.

Kawabata, H. & Zeki, S. (2004). Neural Correlates of Beauty. *Journal of Neurophysiology*, 91, 1699–1705.

Keil, F. C. (2006). Explanation and Understanding. *Annual Review of Psychology*, 57(1), 227–254.

Keil, F. C. & Wilson, R. A., eds. (2000). *Explanation and Cognition*. Cambridge, Mass.: MIT Press.

Kelemen, D. (1999). Function, Goals and Intention: Children's Teleological Reasoning about Objects. *Cognition*, 3(12), 461–468.

Kelemen, D. & Carey, S. (2007). The Essence of Artifacts: Developing the Design Stance. In *Creations of the Mind: Theories of Artifacts and Their Representation*, ed. E. Margolis & S. Laurence (pp. 212–230). Oxford: Oxford University Press.

Kreitler, H. & Kreitler, S. (1972). *The Psychology of the Arts*. Durham, NC: Duke University Press.

Lambie, J. A. & Marcel, A. J. (2002). Consciousness and the Varieties of Emotion Experience: a Theoretical Framework. *Psychological Review*, 109(2), 219–259.

Lamiell, J. T. (1998). "Nomothetic" and "Idiographic": Contrasting Windelband's Understanding with Contemporary Usage. *Theory & Psychology* 8(1): 23–38.

Leder, H., Belke, B., Oeberst, A., & Augustin, D. (2004). a Model of Aesthetic Appreciation and Aesthetic Judgments. *British Journal of Psychology*, 95, 489–508.

Lipton, P. (1991/2004). *Inference to the Best Explanation*. London: Routledge.

Locher, P. J. (2003). An Empirical Investigation of the Visual Rightness Theory of Picture Perception. *Acta Psychologica*, 114(2), 147–164.

Locher, P., Gray, S., & Nodine, C. (1996). The Structural Framework of Pictorial Balance. *Perception*, 25, 1419–1436.

Margolis, J. (1980). Prospects for a Science of Aesthetic Perception. In *Perceiving Artworks*, ed. J. Fisher (pp. 213–39). Philadelphia: Temple University Press.

Margolis, E. and Laurence, S., eds. (2007). *Creations of the Mind: Theories of Artifacts and Their Representation*. Oxford: Oxford University Press.

Martin, L. J. (1906). An Experimental Study of Fechner's Principles in Aesthetics. *Psychological Review*, 13, 142–219.

Martindale, C. (1984). The Pleasures of Thought: A Theory of Cognitive Hedonics. *The Journal of Mind and Behavior*, 5, 49–80.

Martindale, C. (1990). *The Clockwork Muse: The Predictability of Artistic Change*. New York: Basic Books.

Martindale, C., Moore, K., & Borkum, J. (1990). Aesthetic Preference: Anomalous Findings for Berlyne's Psychobiological Theory. *American Journal of Psychology*, 103, 53–80.

Matan, A. & Carey, S. (2001). Developmental Changes within the Core of Artifact Concepts. *Cognition*, 78(1), 1–26.

McManus, I. C., Cheema, B. & Stoker, J. (1993). The Aesthetics of Composition: a Study of Mondrian. *Empirical Studies of the Arts*, 11(2), 83–94.

Miller, J. G. (1984). Culture and Development of Everyday Social Explanation. *Journal of Personality and Social Psychology*, 46, 961–978.

Millikan, R. G. (1984). *Language, Thought, and Other Biological Categories*. Cambridge, Mass.: MIT Press.

Millikan, R. G. (2000). *On Clear and Confused Ideas: An Essay about Substance Concepts*. Cambridge: Cambridge University Press.

Millikan, R. G. (2004). *Varieties of Meaning*. Cambridge, Mass.: MIT Press.

Morris, M. W., Nisbett, R. E., & Peng, K. (1995). Causal Attribution across Domains and Cultures. In *Causal Cognition: A Multidisciplinary Debate*, ed. D. Sperber, D. Premack, & A. J. Premack (pp. 577–613). Oxford: Oxford University Press.

Murphy, G. L. & Medin, D. L. (1985). The Role of Theories in Conceptual Coherence. *Psychological Review*, 92(3), 289–316.

Nichols, S. ed. (2006). *The Architecture of the Imagination: New Essays on Pretence, Possibility, and Fiction*. Oxford: Oxford University Press.

Nichols, S. & Stich, S. (2003). *Mindreading: An Integrated Account of Pretence, Self-Awareness, and Understanding Other Minds*. Oxford: Oxford University Press.

Nisbett, R. E. (2003). *The Geography of Thought: How Asians and Westerners Think Differently ... and Why*. New York: Free Press.

Panofsky, E. (1995). *Three Essays on Style*. Cambridge, Mass.: MIT Press.

Parsons, G. & Carlson, A. (2008). *Functional Beauty*. Oxford: Clarendon Press.

Parsons, M. L. (1987). *How We Understand Art: A Cognitive Developmental Account of Æsthetic Experience*. Cambridge: Cambridge University Press.

Peretz, I. (2006). The Nature of Music from a Biological Perspective. *Cognition*, 100(1), 1–32.

Pickford, R. W. (1972). *Psychology and Visual Aesthetics*. London: Hutchinson.

Pinker, S. (2002). *The Blank Slate: The Modern Denial of Human Nature*. London: Penguin Books.

Ramachandran, V. S. (2001). Sharpening Up "The Science of Art." *Journal of Consciousness Studies*, 8(1), 9–29.

Ramachandran, V. S., & Hirstein, W. (1999). The Science of Art: A Neurological Theory of Aesthetic Experience. *Journal of Consciousness Studies*, 6(6–7), 15–51.

Rhodes, G. (2006). The Evolutionary Psychology of Facial Beauty. *Annual Review of Psychology*, 57(1), 199–226.

Riegl, A. (1890/1992). *Problems of Style*. Princeton: Princeton University Press.

Robinson, J. (1995). Startle. *The Journal of Philosophy*, 92(2), 53–74.

Robinson, J. (2005). *Deeper than Reason: Emotion and Its Role in Literature, Music, and Art*. Oxford: Clarendon Press.

Roskill, M. (1976/1989). *What Is Art History?* 2nd edn [1976]. Amherst: University of Massachusetts Press.

Ruben, D. H. (1990). *Explaining Explanation*. London: Routledge.

Shimamura, A. P. & S.E. Palmer, eds. (2012). *Aesthetic Science: Connecting Minds, Brains, and Experience*. Oxford: Oxford University Press.

Sperber, D. & Wilson, D. (2002). Pragmatics, Modularity and Mind-Reading. *Mind & Language*, 17(1–2), 3–23.

Tanner, J., ed. (2003). *The Sociology of Art: A Reader*. London: Routledge.

Vartanian, O. & Goel, V. (2004). Neuroanatomical Correlates of Aesthetic Preference for Paintings. *NeuroReport*, 15, 893–897.

Walton, K. L. (1973). Pictures and Make-Believe. *The Philosophical Review*, 82(3), 283–319.

Walton, K. L. (1978). Fearing Fictions. *The Journal of Philosophy*, 75(1), 5–27.

Walton, K. L. (1987). Style and the Products and Processes of Art. In *The Concept of Style* (2nd edn), ed. B. Lang (pp. 72–103). Ithaca: Cornell University Press.

Walton, K. L. (1990). *Mimesis as Make-Believe: On the Foundations of the Representational Arts*. Cambridge, Mass.: Harvard University Press.

Watanabe, S., Sakamoto, J., & Wakita, M. (1995). Pigeons' Discrimination of Paintings By Monet and Picasso. *Journal of Experimental Analysis of Behavior*, 63(2), 165–174.

Windelband, W. (1894/1998). History and Natural Science. *History and Theory* 8(1): 5–22.

Wölfflin, H. (1920/1950). *Principles of Art History: The Problem of the Development of Style in Later Art*, trans. M. D. Hottinger. New York: Dover Publications.

Zajonc, R. B. (1980). Feeling and Thinking: Preferences Need No Inferences. *American Psychologist*, 35, 151–175.

Zeki, S. (1999). *Inner Vision: An Exploration of Art and the Brain*. Oxford: Oxford University Press.

13
The Architectonics of the Mind's Eye in the Age of Cognitive Capitalism

Warren Neidich

13.1 Introduction

This essay will use what in 1995 under different political conditions I referred to as a neuroaesthetic argument in order to investigate the recent cognitive turn in cognitive capitalism (Neidich 2013). In the present context I am using this term as it moves beyond its historic meanings embedded as they are in precarious laboring, valorization of capital, and communicative capitalism, and emphasize instead how the knowledge economy and immaterial labor script of a story of neural material engagement made explicit in, for instance, the cooption of neural plasticity and the production of remodeled dynamic neural architectures, so-called neuronal recycling (Dehaene & Cohen 2007). In those first lectures on neuroaesthetics, held at the School of Visual Arts, New York City, I emphasized its anti-cognitivist and ontological conditions in order to de-emphasize its other face as 'neuroesthetics,' in which a neural-based cognitive unfolding of a prescribed a priori genetic inscription determines the perceptual and cognitive abilities of the brain, which then produce its cultural artifacts (Zeki 1999). Instead I sought to emphasize a fluid dialectic relationship between cultural plasticity and neurobiological plasticity, or what I would like to refer to as bidirectional becoming-material engagement, in order to emphasize its dynamic and interactive quality. In this regard neuroaesthetics has adapted itself to the same changing cultural milieu incited by the information economy in which immaterial labor predominates as the dominant form of laboring. A subtext of this essay is that immaterial labor is not so immaterial, because its virtuosic immaterial production leaves traces not as produced objects but as events in the brain's neural circuitry registered as static and distributed neural landscapes. For instance, cognitive capitalism has focused its instrumentalizing dispositifs, such as consumer neuroscience, parametric architecture, cinematic post-production special effects, and social networking codes, directly upon the neural anatomies and neural physiologies of the brain's inherent neural plasticity to shape a conscripted and normalized subject of the future. On the other hand, an activist neuroaesthetic argument utilizes the histories, critiques, practices, apparatuses, spaces, and non-spaces and temporalities of artistic practice as it manifests in painting, sculpture, performance, film, video, and installation art, just to name a few, to counter these arguments and instead incite a different truth production program or alternative paradigm at odds with institutional practices. Twentieth-century art history is full of examples of artistic experimentation from the retinal effects of George Seurat's Pointillist experiments, to the multiple researches of Marcel Duchamp to understand the differences between what he referred to as retinal art and art of the gray matter, to the laboring of Joseph Albers using painting to produce color theory, to the experiments of Op Art artists such as Bridget Riley, who used optical illusions to destabilize subjectivity, to the more recent interest in phenomenology in the works of James Turrell, Olafur Eliasson, and Carsten Höller. I hope to add to these investigations two of my recent artistic experiments – *The Mind's I*, 2009, and *The Noologist's Handbook*, 2012 – in order to show the artistic power of neuroaesthetic methodologies and outcomes. Artistic intrusions do produce alternative and resistant paradigms in the end fabricating artistic facts that are at odds with those brought forth by more mainstream methods. I argue these systems of facts as they coalesce into arguments have the potential to create interventions, which in many cases deterritorialize conventional practices.

Secondly neuroaesthetics is related to what I have termed 'Neuropower' (Neidich 2009, 122; Neidich 2010a, 172; Neidich 2013, 228). Neuropower is an essential component of cognitive capitalism. Neuroaesthetics is one of its agents and generates assemblages of resistant objects through art practices that have adapted themselves to the specific contingencies of cognitive capitalism. As cultural history has demonstrated these objects' powerful material agencies, especially over time, as repositories of a society's and civilization's memory which, as already mentioned, sculpt the brain's neuroplastic potential, producing a so-called dialectic materialism of the neurobiologic substrate. My two performative works *In The Minds' I*, 2008–2011, and *The Noologist's Handbook*, 2011–2013, specifically investigate a second characteristic of 'Neuropower': its action upon the working memory centered in the frontal cortices, through producing internally visualized scenarios that are materialized as concrete objects or displays in a gallery context.

Cognitive capitalism will be defined historically as well as delineating how Taylorist notions of managed efficient workflows of physical movement during industrial labor has transitioned to what I am calling 'Hebbianism' (on which I elaborate below), which constitutes the means though which cognitive labor is made more efficient (Hebb 1949, 62). Devices such as the graphical user interface, where multiple windows appear on the same screen, and the replacement of perspectival space with vernacular space, which is more like cubism than the extended depth of Renaissance perspective, bind components occurring simultaneously together on flat screens as diagrammatic networks. These devices increase worker mental efficiency. This increased efficiency and the attendant attentional relays they facilitate stimulate and wire neurons in activated neural networks more efficiently.

Essential to our understanding of cognitive capitalism is the way in which strategies of mental optimization and the intentionality, specifically its linkage to epistemic ecological niches, have cohered. Epigenetic factors contribute to the ways that these epistemic niches produced especially for cognitive capitalism, call out to the brain's inherent capabilities. 'Epigenesis' is used in the restrictive sense here to refer to the process through which the unfolding of the genetically prescribed brain is altered by its experiences with the environment whether that is the milieu of the brain itself or the world. Neural plasticity describes the degree of anatomic and physiological malleability of neurons, at their synaptic junctions, especially the degree to which they can be modified by experience (Ramachandran & Hirstein 1998, 7; Edelman 2006, 56).[1] While neural plasticity exists in the systems of all animals, it is most prominent in the human cerebral cortex, especially those areas that subserve higher brain functions like language, mathematical ability, and executive functions. Regions of the brain that perform voluntary motor activities and primary sensory functions, like vision and audition, are less malleable (Huttenlocher 2002, 5).

The idea of Material Engagement Theory will be linked to an earlier, "Becoming Cultured-Brain" model in order to introduce a historical model which describes the interactions through time between cognition and material culture (Malafouris 2013; Neidich 2006).[2] In this regard, terms such as parity, complementarity, engram, and exogram will help bridge the intervening space between our internal imagination and the world of externalized and distributed objects to which they are tethered and entangled. Importantly, an alternative definition of exograms as degenerate, polyvalent, and folded complexes in a state of just in time catastrophe will be elaborated. Importantly, the idea of material engagement and cultured brain will initiate a discussion of how these cognitive assists as exogram operate, beyond their object status, as machines or apparatuses that become inculcated in and modulate the operations of the brain itself. This effect of immaterial labor follows those described by Michael Hardt (1994): "Interactive and cybernetic machines become a new prosthesis integrated into our bodies and minds and a lens through which to redefine our bodies and minds themselves."[3] In conclusion, I want to take this statement one step further and look into some of the negative consequences of our contemporary technologies like smart phones and GPS devices acting as cognitive assists upon the brain's substance, in this case the altered topography of the posterior hippocampus, as a model for a generalized understanding of how cognitive capitalism generates a disembodied and normalized subject (Woolett & Maguire 2011).

13.2   Neuropower and Cognitive Capitalism

Cognitive capitalism has two lives. First, emerging from a generalized discussion concerning immaterial labor, it included such ideas as precarity and flexiploitation, being on call 24/7, the socialization of labor such as prosuming and crowdsourcing, becoming profit-rent, capitalization of cognitive labor, and the turn to semiotic or communicative capitalism (Neidich 2012).

Cognitive capitalism is additionally defined by its relation to a cognitive turn in which the brain's neural plasticity and working memory are the new focus of capitalistic adventurism. Neuropower is an essential process of cognitive capitalism. First, neuropower is the latest stage of an ontogenic process, which begins with the disciplinary society, as outlined by Michel Foucault (Deleuze 1992). It was followed by Gilles Deleuze's (1992) society of control as it responded to new forms of technologies of communication like photography, film, telegraphy, and radio, which were immaterial forms of power that were dynamic and no longer respected physical enclosures.[4] Maurizio Lazzarato's (2006) theory of Noo-politics extended the concepts of the society of control to the contemporary conditions of the attention economy, "Noo-politics (the ensemble of techniques of control) is exercised on the brain. It involves above all attention, and is aimed at the control of memory and its virtual power" (Lazzarato 2006, 186). Neuropower is the latest form of biopower and differs from noo-politics in fundamental ways. It is not about the modulation of the attentive networks in the materialized present but is instead about the re-routing of the long- and short-term memories through working memory in the production of future decisions. Second, it acts upon the neural plastic potential of the brain in a living present, especially during what are referred to as the critical periods of development, all the time being guided by the desire to produce a conscripted and enrolled individual of the future. Critical periods are temporal windows in which the nervous system is especially sensitive to the effects of the environment (Neidich 2013). Thirdly, and important for us here in light of the forthcoming experiments I will describe, it redirects the conditions for producing power. In line with neuroscientific research that has exchanged a classical view of the brain as a passive stimulus-driven device that does not actively create meaning by itself but simply reacts to sensory inputs and copies pre-specified information for a more actively engaged model in which internal states are viewed as "action-oriented pointers – that is, as sensorimotor activity patterns that refer to aspects of the world in the context of possible behaviors." (Engel, Fries, & Singer 2001, 704–705).

This new understanding of the brain requires an understanding of top-down processing instead of bottom-up processing. As such, instead of focusing itself upon distributions of sensations which require forms of bottom-up processing – in which abstract concepts are built from concrete sensation – it acts through top-down processing: abstract concepts concerning, for instance, expectation based on experience centered in the forebrain and pre-frontal cortex modulate future actions and behaviors by deciding what information streaming in through downstream sensorial and perceptual systems is salient. This theory of top-down modulation has recently given way to a more distributed approach called a dynamicist view in which neural synchrony and temporal binding play a role. "The model assumes that neural synchrony ... is crucial for object recognition, response selection, attention and sensorimotor integration. Synchrony is supposed to enhance the saliency of neural responses, because correlated discharges have a much stronger impact on neuronal populations than temporally disorganized inputs" (Engel, Fries, & Singer 2001, 706).This description of the mechanisms of top-down control now corresponds to the investigation of the influence of continuously modulating patterns of activity on the processing of sensory signals, and in particular, on their selection and organization through reentrant oscillatory patterning and synchronization. In this complex dynamic model, the patterns relevant to the selection of input signals would be generated not only by assemblies in association cortices that carry more abstract, invariant representations, but as the result of continuous large-scale interactions between higher- and lower-order cortical areas (Engel, Fries, & Singer 2001, 706). Today, it can be advanced that mechanisms or apparatuses of power have increasingly found ways to intervene in this dynamic processing. The working memory is centered in the frontal lobes of the brain and refers to memories held briefly in the mind, making possible the accomplishment of a particular task in the future, and is linked to this dynamic processing. Further results show that "working memory binding involves large-scale neural synchronization at the theta band" (Wu, Chen, Li, Han, & Zhang 2007).

We will see shortly that it is linked to the material world through the relationship between exograms to engrams or what are referred to as brain-artifact interfaces (Malafouris 2010, 264). As such, cognitive capitalism has an effect on the working memory and the image of thought that is constituted there, and as a result, processing of sensory information downstream is important. It is visualized in accordance with specific generationally invented words like rad, dude, networking, and images which are themselves a product of period-specific apparatuses such as parametrically designed buildings and 3-D printed objects as well as social network sites like Facebook and Instagram where members can up load images of themselves called Selfies. The fact is that working memory is under siege in cognitive capitalism. Before continuing our discussion of cognitive capitalism, especially its relation to its extended cognitive turn, we must first become familiar with certain terms such as exogram, engram, brain-artifact interfaces, and the material engagement approach.

13.3   The engram-exographic system and brain artifact interfaces

The viewer is moving in a cinematic space. As a result of the reinvention of space/time coordinates as they become manifest in this new art, language, and architecture, to name just a few, the world in which the spectator and observer moved and lived has changed. It is my opinion that this fundamental change in built space has specific implications for the developing brain. In fact, as these new time-based relations become configured in the physical world, they become remapped into the way neural networks are configured, their spatial relations, and how they operate and communicate to each other through a reorganization of "reentrant temporal flows" (Neidich 2002, 36).

Recently Lambros Malafouris (2013, 37–38) has developed an archeological theory that seeks to describe and explain the long-term relationship between the ontogeny of the interaction between material forms and their processes and the growth and change of human cognitive abilities. The mind can be located within and outside the skin, and human cognition is locationally uncommitted; committed in other words to being uncommitted, distributed, and decentralized. Important for us here and for what is to come is that material engagement takes place along a continuum between theories of internalization and externalization. It is that continuum as it becomes *asymmetric* in contemporary cognitive capitalism as we move into a world of exographic excess that I would like to investigate. However, let me clarify these terms further.

The exogram-engram system is a distributed networked system that does not respect the boundaries of the material world, the body or the brain. Importantly, as we have moved from an extensive, narrative, and linearly mapped world to one that is intensive, non-linear, and self-organized, the nature of engrams and exograms followed suit, mutating separately and together. An engram is a memory record stored in the head. There are at least five dissociable engram or memory systems: 1. Motor skills like writing, driving *and playing video games.* 2. Conditional emotional responses like anxiety created by the sight of a rival *or autistic ones defined by detachment.* 3. Perceptual learning as it relates to learning categories of things like flowers or faces *but also parametrically curvilinear buildings.* 4. Semantic memories that tend to abstract generalizations encoded as language. 5. Episodic memories that relate to the memory of personal experiences in one's life (Donald 2010, 71). Exographic systems have important properties absent in natural memory systems that have implications for human cognition. Examples include totems, masks, knotted cords, built environments, cave paintings, stone circles and burial mounds that operate as astronomical measuring devices, trading tokens, written records, works of poetry, mathematical notations, architectural drawings, libraries and archives, scientific instruments, moving pictures and electronic media, and recently smart phones and robots (Donald 2010, 72).

Essential to any understand of engrams, exograms, brain-artifact interfaces is Clark and Chalmers' 'theory of parity' according to which if part of the world, e.g. a soft-ware program, "functions as a process which were it to go on in the head, we would have no hesitation in accepting it as part of the cognitive process then that part of the world (for that time) is in fact a cognitive process (Clark & Chalmers 1998, 7–19). In other words, portions of the external world can operate as a kind of memory store or exogram, and we can off load operations that would normally go on inside our heads upon external props, thus decreasing the energy requirements of the brain as well as freeing up neural processes to do other things. However, this idea of parity has some problems as it implies that the exogram and the engram are in some way mimetic in their forms, evolution, state relations, and inherent processing operations. Recently the term parity has given way to a theory of complementarity (Malafouris & Renfrew 2010, 7). The term 'complementarity' underscores the lack of exact correspondence between an inner cognitive memory repertoire, engram, and its external cognitive relation, exogram. For instance, "the reformatable nature of exograms allows for information to be altered and then re-entered into storage in ways that an engram clearly can not afford" (Malafouris & Renfrew 2010, 7). In this regard the idea of 'things in motion' as they travel through different epochs and social constructs taking on different meanings and uses is interesting for us here. Furthermore, in order to comprehend the subtleties of the relationships between engram and exogram, as singular entities or as classes of things, it is essential to consider their idiosyncratic diachronic, biographical, and historical aspects (Sutton 2008, 40). Their lack of superimposition, due to a distinctive individual and dyadic character, is related to their inherent developmental asynchronicity and asymmetry. One needs to think of engrams and exograms not as crystallized entities but as intensive interactive folded and plicated membranes. The exograms' polyvalent fields are therefore not equipotential. Their developmental trajectory constitutes a response to the evolving of contextual and contingent cultural tableau that are simultaneously and synchronously emerging and require adaptions to exograms and engrams to counter multiple instabilities that occur in the normal evolution of forms. It is here in their differences that the cognitive life of things, like the cognitive life of brains, can be found where engrams and exograms dynamically start spiking, interacting, and complementing each other (Malafouris and Renfrew 2010, 7).

The theory of the 'becoming-cultured brain' calls for a sympathetic historical materialism of a dynamic and active brain-artifact interface (BAI), which has enabled human beings to further optimize their environments for a more efficient habitation of their world (Malafouris 2010, 265; Neidich forthcoming). In an architectural context BAI is defined as a specified and engineered technological mediation, be it a material structure, process, congregation of objects, socio-material apparatuses, or process, that facilitates the arrangement of a dynamic relationship or tuning between neural and cultural plasticity (Malafouris and Renfrew 2010, 265).[5] Importantly, in cognitive capitalism BAIs are a subset of a whole host of arrangements under the heading of Cognitive Ergonomics through which design platforms optimize cognition-tool interfaces to optimize cognitive laboring (Neidich 2002, 22). I question this univocal concept of BAIs as proposed here through an understanding of the importance of noisy forms in cognitive processes that are negative ergonomic or anti-material. BAIs and the material engagement approach they are imbedded in must be open as the Becoming Cultural Brain model is to the power of noise, chaos, and entropy. For every exogram and engram contains with it unfulfilled promises and possibilities that emerge at points of instability such as in phase changes. It is these instabilities as they morph into singularities that have the potential to disrupt the conditions that create the presentation of the exogram or the engram and push them into morphogenic crisis and that allows them to become other. Two quick illustrations should hopefully suffice: the noise works of John Cage, for instance, his composition 4' 33" in which a piano player sits silently in front of the keyboard doing nothing whilst the sounds generated by an anticipatory crowd generates the noisy, pens dropping, coughing, sound track. In recent times mainstream music is informed by these compositions, and other auto-destructive compositions. On the other hand, one should consider the importance of stochastic resonance. "Recent experiments on single mechanoreceptor neurons of the crayfish show the weak signals can be enhanced by applying an optimal level of external noise ... External noise, in other words, might help the crayfish detect weak signals that otherwise might not" (Kelso 1997, 214).

13.4   From laboring efficient bodies to laboring efficient minds

Recently Malafouris has investigated this concept of brain-artifact interface in order to understand, as is done here, the relationship between cultural plasticity and neural plasticity, which he calls Metaplasticity (Malafouris 2010, 267). Let me summarize his findings before tethering his arguments to those I have previously elaborated in my own theories of cognitive ergonomics, neurobiopolitics, and neuropower (Neidich 2010a; 2013). He divides his analysis into three parts: 1. Mediational effects, which includes enactive prosthetic enhancement and co-evolutionary engagement; 2. Temporal effects; 3. Plastic effects. Prosthetic enhancement enables the mind to utilize or convert the information inscribed in the environment in ways that would otherwise have been impossible for the organism to achieve without such aids. Co-evolutionary material engagement includes human activities that engage with the environment in ways that change it for their more efficient use. In consumer capitalism-designed space, which harbors infinite explicit and implicit clues, such as fields of branded advertisement, which are salient, occurring as they do on multiple platforms such as billboards both imagistic and video, cable television, and computer screens capture attention and produce short-term and long-term memories, in the end producing more diligent consumers. The ominous possibility of epistemic engineering is one possible outcome and was mentioned as the negative consequence of cognitive ergonomics (Malafouris 2010, 266). Temporal binding and anchoring make up the category of so-called temporal effects and are related to the role that is played by BAI in the integration and coordination of processes operating at different time scales. Important in this regard is the notion of binding in the brain. How does the brain know which firing pattern distributed in the multifarious maps across the brain's neural architecture, for instance its visual cortex, basal ganglia, and prefrontal cortex, correspond to which attribute of an object? The visual cortex is divided into various connected modules that code for different attributes of an object like form, color, and movement. Yet we perceive the world as integrated and whole. Binding is responsible for these abilities, and according to "von der Malsburg the brain can tell the neuronal assemblies apart by synchronization. Like the electric lights on the Christmas tree, neurons within one coalition expressing one percept fire in a synchronized manner, but are not synchronized with the coalition coding for another face or for objects in the background" (Koch 2004, 43). Furthermore, in 1990 Francis Crick and Christoph Koch asserted that synchronized 40-hz oscillations with the subset of neurons that correspond to an attended object are a signature of the Neuronal Correlate of Consciousness (Koch 2004, 43). It is no wonder that in consumer capitalism engineered and designed salient objects and fields of salience might be bound by such things as Gestalt properties in order to produce distributions of sensibility with all of its political implications (Rancière 2006, 85). Furthermore, the moving body in space is subject to other forms of rhythmicity caused by itself as it shifts position, as well other moving objects like automobile traffic, other pedestrians, and the swaying of the trees. However, architecture has the potential to produce its own articulatory patterns in order to govern a subject's movement patterns in a specific designed context, and we will look into the notion of the determined diachronic unfolding of movement coordinated by architecture (Schumacher 2012, 34). Finally Malafouris elaborates the plastic effects of BAIs. I will mention these in passing, as much of my later discussion is an intense investigation of these phenomena. Effective connectivity is a form of generational neural plasticity in which cultural memory inscribes itself upon the wet, mutable brain in a lifetime, constituting the ways that BAIs comprise a powerful tool for cultural and activity-based plasticity of the cerebral cortex. The increased size of the gyri of the right motor cortex of a violinist because of the intricate and necessary movements of the left hand as opposed to the right which just moves the bow or the increased size of the representation of the right index finger in the left somatosensory cortex of a right-handed blind braille reader are examples (Pascual-Leone & Torres 1993, 39; Schwenkreis et al. 2007, 3291). Extended reorganization pertains to trans-generational changes in the neural morphology in which the dynamic functional architecture of the cognitive system is changed either by producing new processing nodes or by changing the connections between existing nodes. Stanislas Dehaene's cultural reconversion or 'neuronal recycling hypothesis' in which cultural pressures, for instance language or writing, incite preexisting neural tissue to assume an expanded or different role is a case in point (Dehaene 2004, Dehaene and Cohen 2007).[6]

When cognitive labor becomes the primary form of laboring in the information and knowledge economy, power capitalizes upon these externalized cognitive circuits in order to produce epistemological routines and habits in its subjects. When these publics are working, this yields a way and means with which to optimize their potential labor power by decreasing the systems entropy and thermodynamic waste and yielding greater returns, be they economic, psychological, affective, or social. In Taylorism emphasis on the surveyed and efficient use of physical energies was played out upon the extensive and metric Fordist assembly line. In late Post-Fordism leading to the emergence of cognitive capitalism, these physical exigencies are supplemented through an emphasis on neural efficiencies functioning in the precarious, distributed, and topologic workplace of computational machines. My term Hebbianism is a reference to the famous psychologist D. O. Hebb, who first described the way that synchronously firing neurons that fired together wired together, called Hebb's Postulate: "When an axon of Cell A is near enough to excite Cell B and repeatedly or persistently takes part in firing it, some growth process or metabolic change takes place in one or both cells such that A's efficiency, as one of the cells firing B, is increased" (Hebb 1949, 62). This has significant consequences for us here because Hebb's Postulate has been linked to theories of Long Term Potentiation (LTP) or memory in which persistent changes in synaptic strength is a result of impulse transmission across synapses. Exograms of computational digital interfaces are found inside the computer screen as well as being exported outside into the world of buildings and designed spaces where they have become more prominent in assisting in the articulation of this optimization. They are cognitive artifacts which are cognitive amplifiers, in other words, 'things that make us smart' or maybe dumb (Malafouris 2010, 3). Furthermore, these cognitive artifacts are embedded in the human environment as "small cognitive assists which are drawn on precognitively" (Thrift 2005, 471).[7] Forms of ubiquitous computing and intelligent architecture, which are forms of embedded cognitive assists, create the pleasures of contemporary life, but it also has a darker side. It is in this extended and cognitively embodied digital script that we might discover the brutal and yet delicate relationship between (A)rchitecture as an objective practice, the stuff of the world, and (a)rchitecture as a condition of cerebral substance, what I call Cognitive Architecture (Neidich 2010a). In the past twenty years there has been an excessive production of exograms and the development of what can be called exographic excess. Human experience is increasingly being shaped by exographic medias (Donald 2010, 77). This is leading to a condition I name 'The Incarceration of Contemplation' in which less and less of our time is being spent in deep thinking and more and more of our thinking time is spent wandering the world of material things. (Neidich forthcoming). I develop its consequences in the conclusion to this chapter, but for now will suggest that the art experiments that follow, *In The Mind's I* and *The Noologist's Handbook* can be seen as not only possible forms of resistance but are perhaps new forms of therapy.

The Mind's Eye is the place in the imagination inside your head in which the cinema of the world is projected for detailed inspection. My projects *In The Mind's I* and its later manifestation as *The Noologist's Handbook*, 2008–2011, illustrate the cognitive life of things as well as understanding how various political contexts might affect the conditions of their analogic material re-presentation in the neurobiological architecture of long-term memory as well as how they might be recalled and represented later as real models.

These projects also demand answers to the question of what is an object and how the form of an artistic project may uncover new dimensions for the meaning that objects might hold. *Art before philosophy not after!* It demands to know how the political fantasies of despots, the super abundance of democracy, and the overwrought dialectic of agonism might remake the dynamic constitution of the 'image of thought' that is delineated in the mind's eye.

I would like to take a small detour to describe this project as it was performed recently with students at Southern California Institute of Architecture (SCI-Arc) in Los Angeles and architecture students at the Faculty of Architecture, Ljubljana, Slovenia. It describes how architecture might be used as a 'parallactic cultural device' to hold the image of thought and give it substance as a new form as a neural/architectonic assemblage.

In the summer of 2011 as part of my overall project *University Without Walls* I took architecture students of SCI-Arc on a journey into their imaginations. The curriculum consisted of the following workshops: *Rehearsing the Diagram, Education of the Eye*, and *In the Mind's I*. (Earlier renditions of this project entitled, *In the Mind's I*, were performed in Brussels, Copenhagen, Athens, and Los Angeles.) Each exercised different aspects of the imagination and taught the students how to rethink perception as well how to think/visualize architecture inside their heads, free of the encumbrance of immobile substances and laws of explicit nature like the force of gravity. I might also add that the name for the project was a composite consisting of a reference to 'the mind's eye' as first described by great late 19th and early 20th century psychologist William James as well as a reference to the famous artwork by Robert Morris called the I-Box, 1962, in which a photographic self portrait of the artist's nude body appears behind a cut out capital letter I which like a door swings open revealing the image inlaid in the inside of the container.

The workshop was made up of three parts:

1. Visiting the Schindler House: In the week before the workshop, each student was asked to visit the Schindler House in Los Angeles and to choose one of its rooms to remember later for the performance. This room would become the imagined exhibition space for the work(s) that they would use in their collaboration with me.
2. *In the Mind's I* was performed in a classroom at SCI-Arch and made up of three sections: a. describing the objects; b. imagining the exhibition space; c. collaborating to make the exhibition.
   a. First I asked each student to bring three neutral objects that they were able to carry in one hand from home. We then sat across from each other, upon a specially constructed stage in which the only illumination emanated from a single slide projector that projected colored slides upon us to create shadows that fell on a screen that

separated us from the audience of other students. This was a stage set meant to mimic certain 19th-century shadow plays as well as suggesting the space of Freudian notions of screen memories. The Mind's I performance was meant to redefine these historic agencies and to create a new therapeutic vocabulary for the process of visualization. I then interviewed each student and asked him or her to describe the objects they had brought with them; I had them reveal associated personal memories.

b. Each student was then asked to recall in detail the room they had chosen on their previous trip to the Schindler House. With their eyes closed in front of the audience, I had them describe that room in detail whilst reflecting upon it in their mind's eye. This description delineated for the audience and myself certain knowledge, for instance, of the space's dimensions as well as its lighting conditions and layout, paying attention to details like fireplaces and windows.

c. Subsequently, I asked each to collaborate with me to create an imaginary exhibition, with the objects and stories that they had brought with them, in that make-believe space that they conjured in their mind's eyes. Together we created about three to four imaginary works of art. We then installed the works there, specifying, for example, their location, lighting and distance from each other. In my earlier renditions the work had ended here resulting in a purely immaterial and imaginary artwork.

3. However, two weeks later I added another component. I arrived to class and asked the students to make architectural models of their imaginary works. This led to the final component of the work entitled The Production of the Model. Based on their memories of their exhibitions each was successful in producing their 'replica' from memory. The next week they exhibited those models and talked about how they related to their original performance with special emphasis on the difference between how they remembered it, and how, in fact, the model represented it.

13.5   *In the Mind's I* becomes *The Noologist's Handbook*, at P74 Gallery in Ljubljana

What is a noologist? It is an artist/curator who transforms the image of thought. In Ljubljana, Slovenia, another component was added to the *In the Mind's I* project. As *In the Mind's I* project architectural students again brought three objects with them to the encounter, again they envisioned an exhibition space in their mind's eye. In the third stage a radical shift was instituted that affected the performance in dramatic ways. Each curator was assigned a specific character out of which emerged a specific curatorial strategy. These were the generous, agonistic, and despotic curators. Any time an artist asked to do something in their work, the despotic curator responded no and rejected it. The despotic curator sculpted the exhibition according to his or her own desire, not leaving any room for the wishes of the artist. The generous curator was just the opposite, saying yes to every suggestion and at times acting as a cheerleader. This role-playing had effects not only on the kind of imaginary art work the artist concocted and which was later produced for the P74 Gallery but also how the audience interpreted and visualized the exhibition. In Ljubljana members of the audience, referred to as the 'mind's we', were given paper and pencil, and when possible they were asked to draw, during and after each performance, the imaginary exhibition they envisioned. These drawings were collected and displayed on the wall at P74 Gallery and commented upon. The agonistic situation was difficult to analyze because the complexity of the conversation made it not only difficult to draw but also to later interpret. Even though these were exhibited, they were not included in the final analysis. For instance, the despotic drawings had the picture of the dominatrix in many of the pictures, one image was drawn in perspective, itself a kind mental incarceration, the drawings were made with heavy dark lines which were drawn at very acute spiky angles while the images generated from the generous curator were light, dreamy, filled the entire page and had images of happy faces. Thus the performance became metonymic or a stand-in for how political power has the potential to modulate the kinds of images circulating through the mind's eye and the mind's we. Of course we assume this, but the audience drawings were a kind of physical proof. The despotic curator stood in for the despotic politician. In tertiary economies in which cognitive capitalism is the new regime, sovereignty, using its own apparatuses, constitutes the mind's eyes of its subjects and can open up or restrict the image of thought.

13.6   The materiality of immaterial labor

At this point, a digression to further analyze the evolving conditions of new labor and their implications for the ontogeny of the machinery of the brain and mind is necessary. Paolo Virno suggests in *A Grammar of the Multitude* that in the information economy, the communicative act itself functions as an attractor repositioning political action, labor, and intellectual reflection closer to one another:

Let us consider carefully what defines the activity of virtuosos, of performing artists. First of all, theirs is an activity which finds its own fulfillment...in itself, without objectifying itself into an end product, without settling into a 'finished product' or into an object which would survive the performance. Secondly, it is an activity which requires the presence of others, which exists only in the presence of an audience (Virno 2004: 50).

Furthermore, I will maintain in particular, that the world of so-called post-Fordist labor has absorbed into itself many of the typical characteristics of political action: and that this fusion between politics and labor constitutes a decisive physiognomic trait of the contemporary multitude. Notice how this reference to physiognomy underscores the multitude's recognizable faciality, its ability to be dominated by the sovereign who now can shape face according to a template of for instance the Caucasian male, and, secondly, owing to the sophisticated technologies at hand, sovereignty is today able to judge and normalize the multitude with the help of sophisticated apparatuses such as consumer neuroscience, big data, and software agents, preempting our feelings, tracking and analyzing our every click.

Beginning with Aristotle's *Nicomachean Ethics*, Virno delineates the differences between what he refers to as labor (or poesis) and political action (praxis) through the idea of virtuosity, distinguishing the former from the latter by the production of an object, in the case of poesis, and the lack of one, in the case of praxis. In praxis the purpose of the action is found in the action itself! Later, expanding his argument through Marx's discussions in *Results of the Immediate Process of Production* and *Theories of Surplus Value*, he describes two kinds of intellectual labor. First, he characterizes a form that produces products of mental activity like books and paintings as well as instances in which the products and acts of producing them are inseparable from the act itself. Secondly, he specifies types of labor such as a virtuosic performance that leaves no real product or trace. Here, he includes pianists, dancers, orators, teachers, and even butlers (Virno 2004, 56). Would he include my Noologist as well? Yet, unless that speaker is speaking to himself or herself, an act considered somewhat odd and demented when it addresses to no one yet occurs in the context of others and which is therefore deflected by possible recipients as nonsense, or self-reflexive when it is part of a singular and lone rehearsal, his or her speech finds a receptor-listener. How essential this is to recent theories of mind, referring as it does to our ability to form an insight into what other people are thinking in order to anticipate their behavior. Deception is an important attribute of such a theory in which the deceiver manipulates the mental states of another person in order to later exploit them. When that listener is not one but many, every speech act finds, creates, or produces an audience. This need for an audience or a collective mind as a roving, wet, mutable organic interface, where the inscription of the oral history/memory of that performance is inscribed in both the static and dynamic conditions of the material brain, figures the central key to what follows. The virtuoso performance does in fact produce a material change and, therefore, can leave a trace as mental sculptures and architectures. Production of such traces shapes the essence of neoliberal capitalism and new labor. When the agency of this neoliberalism focuses directly upon the conditions of cognition in all its variety, a set of conditions arise that I shall call neoliberal cognitive capitalism. I argue that the essence of new labor acting in concert with the political, social, and cultural habitus with which it interacts, produces neural efficiencies both local and global, micro and molar, occurring both in the world as designed space and architecture as well in the stable and dynamic potentialities of the circuitry of the brain. In the end there is an ontogenic interaction, both in the external material life of the individual, his world of things and the life inside, between the functional integrity of the world, how the relations of the world are organized together as evolving assemblages of multiple coexisting templates waiting just in time to be acknowledged, and their counterpart, developmentally organized functionally integrated dispersed assemblies in the brain.

In the Mind's I reveals that something is, in fact, left behind as a trace, and this suggests a new kind of materialism that is essential to the delineation and unveiling of the conditions becoming ever more pervasive in cognitive capitalism: one that leaves residual traces in the neurobiological matrix in the mind of the collaborating artist. The virtuoso performance produces a material change as 'mental memory sculptures' and architectures manifesting themselves as subtle stabilizations and destabilizations of the structural and dynamic neuro-biochemical conditions of the brain. What the artist and audience listening to my performance 'take home with them' is not a real object or artwork but an immaterial and imaginary one: an engram that is accessible later, as we saw with the architects as SCI-Arch who were able to re-constitute their remembered performative models. However, in the second performative experiment *The Noologist's Handbook*, we were able to witness another level to this materiality thanks to two events that occurred. First, the establishment of the concept of In the Mind's We. Also, the audience first witnessed and then materialized the contents of their 'mind's we' as drawings were made during each of the performances. Further, something else was happening which was interesting for the establishment of theories of cognitive capitalism as the place for the production of a normalized and subtle subject. Each of the curatorial personas produced different forms in the theater of the imagination. The despotic curator produced exhibitions that were hard-edged, focused, Cartesian and contained images of a dominatrix character, while the generous curator produced images quite differently characterized. The drawings produced under the generous curator were fully inscribed pages that went all the way to the edges, images of happy faced children and light markings. These were qualitative judgments made by artists who not only attended the exhibition, but also made drawings hung as art on the wall, and then also became the judges of the art. Thus the idea of the individual and the audience underwent transformation.

13.7 Conclusion

> He who lets himself be captured by the 'cellular telephone' apparatus – whatever the intensity of the desire that has driven him – cannot acquire a new subjectivity, but only a number through which he can, eventually, be controlled. The spectator who spends his evenings in front of the television set only gets, in exchange for his desubjectification, the frustrated mask of the couch potato, or his inclusion in the calculation of the viewership ratings (Agamben 2009, 21).

I use this conclusion to understand the politics of a recent trend in cognitive capitalism: the current excess of cognitive assisting devices sold as consumer products. These inventions, such as smart phones and GPS devices, have become part of our daily life and have continued a trend already in place instigated by consumer culture: the asymmetry of material engagement in which the normal relations between thought processes going on inside one's head and outside becomes imbalanced. I am referring to this capturing and externalization of thought at the expense of deep internalized contemplation as 'the incarceration of contemplation'. Just for a moment think of how much time one spends working on computers, using one's smart phone global positioning systems for destination location, and using social networking sites as a substitute for real voice communication. Earlier in this essay this condition was described as a diminution of time spent in deep thought, a loss of epistemological coordination in prognosticating, a decrease in the ability to analyze complex political situations leading to a supple and easily normalized subject, a decrease in the importance of the contents of one's private life and their outing in Facebook and other forms of social media as well as denigration of talk therapy in favor of pharmaceuticals.

I would like to refer back to the subtle relationship already developed earlier between noopolitics and neuropower to further investigate further 'the incarceration of contemplation'. Recall that noopolitics was described as the ensemble of techniques of control that above all involves attention and is aimed at one's memory systems (Lazzarato 2006). Although memory is a complex entity, at least its short-term acquisition, it is intimately involved with attention. "Attention, rehearsal, repetition and practice are cognitive operations that work synergistically in the making or strengthening of the synapses that form the memory networks of the cortex" (Fuster 2003, 115). I suggest that this mental outsourcing produces a dangerous situation for the body on two fronts. First, the corporeal body in its disengagement is left defenseless and subject to motor vehicle accidents. Second, the mind, which in spite of the obvious positive side effects of GPS devices of spatial knowledge and location awareness, is subject to the effects of disuse.

Recently the Los Angeles Department of Transportation announced steps to combat a rise in pedestrian deaths due to texting while crossing the street. The term distracted walking has been used to define this condition. "More than 1,500 pedestrians were treated in emergency rooms in 2011 after being injured while using a portable electronic device such as a cellphone, according to a recent U.S. Consumer Product Safety Commission Report" (Villeneuve 2013). Using the model first introduced by Paul Virilio (1994, 14) of the phatic image, "phatic image – a targeted image that forces you to look and holds your attention," the LED screen of the smart phone is a kind of super phatic image that calls out to our attention and captures it so completely that even the normal attention grabbing system of signs and symbols of contemporary designed urban space cannot compete with it (Virilio 1994). Smart phone cognitive ergonomics trumps urban semiotics and cognitive mapping to the dismay of those in charge of keeping the public safe.

Secondly, recent research concerning the relationship of learning the streets of London, the so-called 'Knowledge', by student cab drivers, and the consequent hypertrophy and atrophy of the hippocampus illustrates the neuroplastic potential of contemporary cognitive assists and exograms as neuromodulators. The works of the neuroscientists

Katherine Woollett and Eleanor A. Maguire are central here. They found an increase in the size of the posterior hippocampus, where new short-term memories are stored, between those taxi drivers studying for their drivers' licenses and those not. A reversal of these findings or atrophy of the hippocampus resulted in retired taxi-cab drivers (Woollett & Maguire 2011). Furthermore, almost half of those taking the course quit, suggesting that gene polymorphisms might lead to different subject capabilities in learning map reading related to driving (Woollett & Maguire 2011). Thus, the trainees who finished the course might have had a genetic predisposition that enhanced their ability to learn in this context (Deacon 2001, 100). Finally the work of Veronique Bohbat at Toronto Hospital Canada found that the reliance of GPS devices by normal drivers may reduce hippocampal function (Edwards 2010). As such, learning of mapping of the urban space with cognitive assists devices can reduce brain function. Here is the central hub of the network of ideas that constitute this thesis. In conclusion, the politics of cognitive capitalism is linked to a series of computational devices and spatial assists that produce neuromodulatory consequences, which can expand or contract the power of the mind's eye.

Notes

1. It is a characteristic of neural tissue more early in life, during what are described as critical or sensitive periods, than later and can operate at the granular-cellular level, at what are called post-synaptic membranes as well as more molar-network level such as in cases of remapping where cortical maps in the somesthetic cortex become reorganized in response to loss of a limb.
2. "In a brain that has been selected for through the operation of neural Darwinistic and neural constructivist pressures, the spatial configurations of neurons and networks and their non-linear, dynamic neural signatures manifest as synchronous oscillatory potentials; they reflect the influence of this complex, competing, artificially created network of phatic signifiers that dominate the contemporary visual landscape. Drawing attention to these processes of binding and dispersal, I propose that as the systems of technical/cultural mediation become increasingly more folded, rhizomatic and cognitively ergonomic, they evolve to more closely approximate the conditions of temporal transaction that sculpt the intensive brain" (Neidich 2006, 222).
3. Important for my argument here is the following passage: "Today, as general social knowledge becomes ever more a direct force of production, we increasingly think like computers, and the interactive model of communication technologies becomes more and more central to our laboring activities. Interactive and cybernetic machines become a new prosthesis integrated into our bodies and minds and a lens through which to redefine our bodies and minds themselves" (Hardt 1995, 94–95). For further discussion of these issues (from social brain to affect) see Wolfe 2010 and Wolfe forthcoming.

4. The passage from the disciplinary society to the society of control and noo-politics, that is to say the administration in the closed and wide-open spaces, previously focused on the condition of the individual and the dividual in relation to the past and the present. They described the focus of power as that which organized the interruptions and undulation of flows of time and space in the disciplinary society and society of control, respectively, in the context of the present now, even if, for instance, in the society of control Deleuze suggests future kinds of gadgets of control, such as an "electronic card that raises a given barrier" (Deleuze 1992, 7).

5. These sort of bidirectional dynamic coalitions that lie at the heart of BAIs can take many forms (eg. hard assembled (stable)/soft assembled (reconfigurable) epistemic/pragmatic, invasive/non-invasive, representational performative, transparent/non-transparent, constitutive/instrumental, etc.) and can be empirically observed through diverse examples ranging from the early stone tools to the more recent symbolic technologies such as calendars, writing, and numerals as well as pencils and paper. One could add that brain machine interfaces (BMIs) make it now possible for a monkey or human to operate remote devices directly via neural activity.

6. "1. Human brain organization is subject to strong anatomical and connec-tional constraints inherited from evolution. Organized neural maps are present early on in infancy and bias subsequent learning.2. Cultural acquisi-tions (e.g., reading) must find their 'neuronal niche,' a set of circuits that are sufficiently close to the required function and sufficiently plastic as to reorient a significant fraction of their neural resources to this novel use.3. As cortical territories dedicated to evolutionarily older functions are invaded by novel cultural objects, their prior organization is never entirely erased. Thus, prior neural constraints exert a powerful influence on cultural acquisition and adult organization" (Dehaene 2007, 384–385).

7. This understated building presages and /or duplicates the world by anchoring more and more of what was regarded as 'human' in the 'environment' in the form of small cognitive assists which are drawn on precognitively.

References

Agamben, G. (2009). *What Is an Apparatus? And Other Essays*, trans. D. Kishik and S. Pedatella. Palo Alto: Stanford.

Clark, A. & Chalmers, D. (1998). The Extended Mind. *Analysis* 58, 7–19.

Deacon, T. (2001). Multilevel Selection and Language Evolution. In B. H. Weber & D. J. Depew (eds), *Evolution and Learning: The Baldwin Effect Reconsidered (Life and Mind)* (pp. 81–107). Cambridge, Mass.: MIT Press.

Dehaene, S. (2004) Evolution of Human Cortical Circuits for Reading and Arithmetic: The "Neuronal Recycling" Hypothesis. In S. Dehaene, J. R. Duhamel, M. D. Hauser, & G. Rizzolatti, *From Monkey Brain to Human Brain. A Fyssen Foundation Symposium.* Cambridge, Mass.: MIT Press, viewed August 14, 2013, http://citeseerx.ist.psu.edu/ viewdoc/download?doi=10.1.1.57.9491 &rep=rep1&type=pdf.

Dehaene, S. & Cohen, L. (2007). Cultural Recycling of Cortical Maps. *Neuron*, 56, 384–398.

De Boever, A. & Neidich, W., eds. (2013). *The Psychopathologies of Cognitive Capitalism*. Berlin: Archive Books.

Deleuze, G. (1992). Postscript on the Societies of Control. *October*, 59, 3–7.

Donald, M. (2010). The Exographic Revolution: Neuropsychologial Sequelae. In L. Malafouris & C. Renfrew (eds), *The Cognitive Life of Things: Recasting the Boundaries of the Mind* (pp. 71–79). Cambridge: McDonald Institute of Monographs.

Edelman, G. M. (2006). *Second Nature*. New Haven and London: Yale University Press.

Edwards, L. (2010). Study Suggests Reliance on GPS May Reduce Hippocampus Function as We Age. Phys Org, http://phys.org/, viewed August 1, 2013, http://phys.org/news/2010-11-reliance-gps-hippocampus-function-age.html

Engel, A. K., Fries, P., & Singer, W. (2001). Dynamic Predictions: Oscillations and Synchrony in Top-Down Processing. *Nature Reviews Neuroscience*, 2, 704–716.

Fuster, J. M. (2003) *Cortex and Mind*. New York: Oxford University Press.

Hardt, M. (1999). Affective Labor. *Boundary 2*, 26(2), 89–100.

Hebb, D. O. (1949). *The Organization of Behavior*. New York: Wiley.

Huttenlocher, P. R. (2002). *Neural Plasticity*. London: Harvard University Press.

Kelso, J. A. Scott (1997). *Dynamic Patterns*. Cambridge, Mass.: The MIT Press.

Koch, C. (2004). *The Quest for Consciousness*. Englewood: Roberts and Company Publishers.

Lazzarato, M. (2006). Life and the Living in the Societies of Control. In M. Fuglsang & B. M. Sorensen (eds), *Deleuze and the Social* (pp. 171–190). Edinburgh: Edinburgh University Press.

Malafouris, L. (2010). The Brain–Artefact Interface (BAI): A Challenge for Archaeology and Cultural Neuroscience. *Social Cognitive and Affective Neuroscience*, 5(2–3), 264–273.

Malafouris, L. (2013). *How Things Shape the Mind*. Cambridge, Mass.: The MIT Press.

Malafouris, L. & Renfrew, C. (2010). The Cognitive Life of Things: Archeology, Material Engagement and the Extended Mind. In L. Malafouris & C. Renfrew (eds), *The Cognitive Life of Things: Recasting the Boundaries of the Mind* (pp. 1–12). Cambridge, UK: McDonald Institute of Monographs.

Neidich, W. (2002). *Blow-up: Photography, Cinema and the Brain*. New York: Distributed Art Publishers.

Neidich, W. (2006). The Neuro-biopolitics of Global Consciousness. *Sarai Reader 06: Turbulence*, 222–236.

Neidich, W. (2009). Neuropower. *Atlántica Magazine of Art and Thought*, 48–49, 118–167.

Neidich, W. (2010a). From Noopower to Neuropower: How Mind Becomes Matter. In D. Hauptmann & W. Neidich (eds), *Cognitive Architecture: From Bio-politics to Noo-politics*. Rotterdam: 010 Publishers.

Neidich, W. (2010b). Sculpting the Brain and I don't Mean Like Rodin. In W. Neidich & R. Premath (eds), *Shifter: special edition on Pluripotential*, 16, 173–187.

Neidich, W. (2012), The Artist Residency in the 21st Century: Experiments in Cultural Potentiality and Contamination. *ArteEast Quarterly Magazine*, viewed August 14, http://www.arteeast.org/2012/02/20/761/.

Neidich, W. (2013). Neuropower: Art in the Age of Cognitive Capitalism. In A. DeBoever and W. Neidich (eds), *The Psychopathologies of Cognitive Capitalism* (pp. 219–261). Berlin: Archive Books.

Neidich, W. (forthcoming). Computation Architecture in the Age of Cognitive Capitalism. In Liss Werner, ed., *Encoding Architecture*. Pittsburgh: Carnegie Mellon University Press.

Neidich, W., ed (2013). *The Psychopathologies of Cognitive Capitalism: Part 2* (forthcoming). Berlin: Archive Press.

Pascual-Leone, A. & Torres, F. (1993). Plasticity of the sensorimotor cortex representation of the reading finger in Braille readers. *Brain*, 116, 39–52.

Rancière, J. (2006). *The Politics of Aesthetics*, trans. G. Rockhill. London: Continuum.

Ramachandran, V.S. & Hirstein, W. (1998). The Perception of Phantom Limbs. The D.O. Hebb lecture. *Brain*, 121, 1603–1630.

Schumacher, P. (2012). *The Autopoiesis of Architecture*. West Sussex: Wiley.

Schwenkreis, P., El Tom, S., Ragert, P., Tegenthoff, M., & Dinse, H. R. (2007). Assessment of Sensorimotor Cortical Representation Asymmetries and Motor Skills in Violin Players. *European Journal of Neuroscience*, 26, 3291–3302.

Sutton, J. (2008). Material Agency, Skills and History: Distributed Cognition and the Archeology of Memory. In C. Knappet & L. Malafouris (eds), *Material Agency, Towards a Non-Anthropocentric Approach* (pp. 37–55). New York: Springer.

Thrift, N. (2005). From Born to Made, Technology, Biology and Space. *Transactions of the Institute of British Geographers*, 30, 463–476.

Villeneuve, M. (2013). Texting on Your Feet Can Be a Safety Hazard. *L.A. Times*, 6 August, A6.

Virilio, P. (1994). *The Vision Machine*, trans. J. Rose. Bloomington: Indiana University Press.

Virno, P. (2004). *A Grammar of the Multitude*, trans. I. Bertoletti, J. Cascaito, and A. Casson. Los Angeles: Semiotext(e).

Wolfe, C. T. (2010). From Spinoza to the Socialist Cortex: Steps Toward the Social Brain. In *Cognitive Architecture. From Bio-politics to Noo-politics*, eds. D. Hauptmann and W. Neidich (pp. 184–206). Rotterdam: 010 Publishers, Delft School of Design Series.

Wolfe, C. T. (forthcoming). Cultured Brains and the Production of Subjectivity: The Politics of Affect(s) as an Unfinished Project. In *The Psychopathologies of Cognitive Capitalism II*, eds. M. Pasquinelli and W. Neidich. Berlin: ArchiveBooks.

Woollett, K. & Maguire, E. (2011). Acquiring "the Knowledge" of London's Layout Drives Structural Brain Changes. *Current Biology*, 21, 2109–2114.

Wu, X., Chen, X., Li, Z., Han, S., & Zhang, D. (2007). Binding of Verbal and Spatial Information in Human Working Memory Involves Large-scale Neural Synchronization at the Theta Frequency. *Neuroimage*, 35, 1654–1662.

Zeki, S. (1999). Art and the Brain. *Journal of Consciousness Studies*, 6, 76–96.

Index